REMOVAL OF METALS FROM WASTEWATER

REMOVAL OF METALS FROM WASTEWATER

Neutralization and Precipitation

Edited by
George C. Cushnie, Jr.
CENTEC Corporation
Reston, Virginia

Prepared by
A.S. Goldfarb, C.V. Fong, W. Lowenbach, J. Schlesinger
The MITRE Corporation
McLean, Virginia

W.A. Parsons, S.J. Vyas
Davy McKee Corporation
Cleveland, Ohio

P.A. Vesilind
Duke University
Durham, North Carolina

NOYES PUBLICATIONS
Park Ridge, New Jersey, USA

Copyright © 1984 by Noyes Publications

Library of Congress Catalog Card Number: 83-22142
ISBN: 0-8155-0976-6
ISSN: 0090-516X
Printed in the United States

Published in the United States of America by
Noyes Publications
Mill Road, Park Ridge, New Jersey 07656

10 9 8 7 6 5 4 3 2 1

Library of Congress Cataloging in Publication Data

Main entry under title:

Removal of metals from wastewater.

   (Pollution technology review ; no. 107)
   Bibliography: p.
   Includes index.
   1. Sewage--Purification--Heavy metals removal--
Handbooks, manuals, etc. 2. Sewage--Purification--
Neutralization--Handbooks, manuals, etc. 3. Sewage
--Purification--Precipitation--Handbooks, manuals, etc.
I. Cushnie, G.C. II. Goldfarb, Alan S. III. Series.
TD758.5.H43R46 1984   628.3'57   83-22142
ISBN 0-8155-0976-6

Foreword

This book is a manual of design and operating procedures for the removal of metals from industrial wastewaters by neutralization and precipitation. Also covered are methods for handling and disposal of residues from the treatment processes.

Effluent limitations for point source discharges into waterways or publicly owned treatment works have placed particular emphasis on the control of toxic materials. Among the toxic materials identified by the U.S. Environmental Protection Agency are the heavy metals including beryllium, cadmium, chromium, copper, lead, mercury, nickel, silver, thallium, and zinc. This manual is designed as a practical guide to the management of a wastewater treatment program for those involved with the removal of metals from wastewaters.

The book contains reviews of legal requirements, wastewater management practices, fundamental process chemistry, engineering design concepts and procedures, cost estimation, the state of the art of control of precipitate properties, and methods used for sludge treatment and disposal. It is a comprehensive source of practical information for the plant manager or operator, the engineer or consultant seeking to prepare preliminary process designs and cost estimates or to evaluate proposed treatment systems. Numerous graphical illustrations and an appendix of sample problems are included to enhance the reader's understanding of the material presented.

The information in the book is from *Manual of Practice for Wastewater Neutralization and Precipitation,* prepared by A.S. Goldfarb, C.V. Fong, W. Lowenbach, and J. Schlesinger of the MITRE Corporation; W.A. Parsons and S.J. Vyas of Davy McKee Corporation; and P.A. Vesilind of Duke University; and edited by G.C. Cushnie, Jr. of CENTEC Corporation for the U.S. Environmental Protection Agency, April 1981.

The table of contents is organized in such a way as to serve as a subject index and provides easy access to the information contained in the book.

>Advanced composition and production methods developed by Noyes Data are employed to bring this durably bound book to you in a minimum of time. Special techniques are used to close the gap between "manuscript" and "completed book." In order to keep the price of the book to a reasonable level, it has been partially reproduced by photo-offset directly from the original report and the cost saving passed on to the reader. Due to this method of publishing, certain portions of the book may be less legible than desired.

Acknowledgments

The contributions of Mr. Fred Ellerbusch and Mr. Joginder Bhutani, former MITRE employees, and Mr. Paul N. Cheremisinoff, President of Pollution Engineering Technology, Inc., are gratefully acknowledged. Special thanks are extended to Dr. Asad T. Amr, Dr. Ebenezer Arkene-Afful, Dr. Elbert C. Herrick, and Dr. Paul R. Clifford for their careful review of and constructive comments on the draft version of the manuscript.

We are particularly grateful to the late Mr. George F. Weesner, who, as project officer, provided encouragement and many helpful suggestions on the preparation of this manual.

NOTICE

This report has been reviewed by the Industrial Environmental Research Laboratory, U.S. Environmental Protection Agency, and approved for publication. Approval does not signify that the contents necessarily reflect the views and policies of the U.S. Environmental Protection Agency or the publisher, nor does mention of trade names or commercial products constitute endorsement or recommendation for use.

Contents and Subject Index

1. INTRODUCTION 1
2. PREDESIGN CONSIDERATIONS 4
    Water Use Survey 4
    Wastewater Monitoring. 6
        Wastewater Characteristics 10
            What to Measure 10
            When to Measure. 10
            Where to Sample 11
        Data Analysis 13
    Methods of Wastewater Reduction 15
        Flow Reduction 15
        Spill Prevention and Control 17
        Countercurrent Operations 19
        Flow Equalization. 22
            Batch Processing 22
            Continuous Processing from Batch Storage. 24
            Side-Stream Equalization 24
            Flow-Through Equalization. 24
        Stream Segregation 26
        Water Reuse. 28
    Laboratory Evaluation 29
        Methods 29
        Reagent Selection 30
    Summary. 32
3. PROCESS DEVELOPMENT 34
    Selection of Reagent 34
        Ammonia. 38

Carbon Dioxide. 38
Lime Materials 38
    Quicklime 39
    Hydrated Lime 39
Sodium Carbonate. 40
Sodium Hydroxide 40
Sodium Metabisulfite. 40
Sodium Sulfide. 40
Sulfur Dioxide 41
Sulfuric Acid 41
Determination of Reagent Quantity .41
Development of the Rate Equation. 47
Differential Analysis 47
Integral Analysis 49
Statistical Analysis. 50
Control of Precipitate Properties 52
Batch Neutralization 52
Continuous Flow Neutralization. .53

4. PROCESS DESIGN FOR NEUTRALIZATION AND PRECIPITATION 55

Reactor Sizing and Geometry 55
Mixing System Design 58
Energy Requirements. 58
Dead Time. 59
Scale-Up 62
Flocculation. 65
Control System Design. 65
Range of Operation. 66
Attenuation 67
Aspects of Reactor Design. 67
Reactor Staging. 72
Graphical Back-Mix Design Procedure 73
Performance Projection 76
Back-Mix Reactor Design Theory 78
    Variable Feed. 80
    Variable Concentration. 81
    Staging Control. 81
Control of Crystalline Properties 83
    Experimental Procedure 84
    Practical Application 87
Chemical Handling and Storage 89
Liquid Reagents 89
Gaseous Reagents 91
Polyelectrolytes 91
Construction Materials 92
Masonry 94
Metals. 95

| Plastics 96
| Elastomers 97

5. COST ESTIMATION 98
Mixed Reactor Costs 99
Construction 99
Operation and Maintenance 99
Sodium Hydroxide Feed System Costs 99
Construction 99
Operation and Maintenance 101
Hydrated Lime Feed System Costs 101
Construction 101
Operation and Maintenance 103
Ammonia Feed Facilities 106
Anhydrous Ammonia Construction Costs 106
Anhydrous Ammonia Operation and Maintenance Costs 106
Alum Feed System Costs 109
Construction 109
   Liquid Alum 109
   Dry Alum 110
Operation and Maintenance 110
Polymer Feed System Costs 113
Construction 113
Operation and Maintenance 114
Flocculation System Costs 116
Construction 116
Operation and Maintenance 116
Other Costs 117

6. SUSPENDED SOLIDS SEPARATION 122
Sedimentation 127
Classes of Particle Settling 127
   Class 1–Free Settling 129
   Class 2–Hindered Settling 130
   Classes 3 and 4–Zone Settling and Compression Settling ... 132
Aids to Settling 136
Design 136
   Scale-Up 136
   Short Circuiting 137
   Turbulence 138
   Bottom Scour Velocity 138
   Inlet Design 138
   Basin Depth 139
   Outlet Design 141
Equipment and Operation 141
Sizing 143
Economics 148
Flotation 149

    Design..153
    Equipment.....................................155
    Operation.....................................156
    Sizing..156
    Economics.....................................158

7. SLUDGE DEWATERING AND DISPOSAL..................160
Sludge Dewatering Techniques......................161
  Centrifugation..................................161
    Operation.....................................161
    Sizing..166
    Economics.....................................170
  Filtration......................................171
    Operation.....................................171
    Sizing..173
    Economics.....................................177
  Drying..177
    Operation of Sand Drying Beds.................178
    Sizing of Sand Drying Beds....................180
    Economics of Sand Drying Beds.................180
Sludge Disposal..................................180
  Landfill..181
  Landfills With Encapsulation and Fixation.......182
  Deep Well Injection.............................183
  Contract Disposal...............................184

REFERENCES..185

BIBLIOGRAPHY......................................193

APPENDIX—SAMPLE PROBLEMS..........................200
Example 1..200
  Problem...200
  Solution..201
Example 2..202
  Flow Equalization—Graphical Procedure...........202
  Calculation of Minimal Reservoir Volume Required to Operate
  Continuously at Average Flow Rate...............204
Example 3..207
  Problem...207
  Solutions.......................................209
    Determination of Reaction Rate Equation.......209
      Differential Analysis.....................209
      Statistical Analysis......................213
Example 4..216
Example 5..217
Example 6..219
  Given...219

Contents and Subject Index   xiii

    Required 221
        Scale-Up 222
        Reactor Dead Time 223
    Discussion 224
Example 7 225
    Given 225
    Solution 225
Example 8 228
    Given 228
    Required 228
    Calculate Annualized Costs 229
Example 9 231
    Given 231
    Solution 231

1
Introduction

The 1972 and 1977 amendments to the Federal Water Pollution Control Act (Public Law 92-500) require the U.S. Environmental Protection Agency (EPA) to establish effluent limitations for point source discharges into waterways or publicly owned treatment works. The 1977 amendments place particular emphasis on the control of aqueous discharges of toxic materials. Among the toxic materials identified by EPA are the heavy metals, which include beryllium, cadmium, chromium, copper, lead, mercury, nickel, silver, thallium, and zinc. Under the law, all industries must meet Best Available Technology (BAT) standards for toxic discharges by July 1, 1984. Figure 1 summarizes the activities that may be required of plant owners and operators facing compliance with EPA regulations regarding the discharge of metals from their plants.

This manual is designed to provide practical information to plant owners and operators, engineers, consultants, and regulatory officials regarding the management of a wastewater treatment program and the design of a treatment system for the removal of metals from wastewater. The process of neutralization and precipitation is the state-of-the-art technique for removing metals from wastewater. The information contained in this manual is directed toward design and operating procedures for the neutralization and precipitation process and the handling and disposal of the residues from the process.

This manual was written for a wide audience with varying interests and needs. Each section contains material of interest to a different segment of the audience. Section 2 provides information on: techniques for reducing the wastewater load, methods for determining the plant wastewater characteristics to establish the design basis for a wastewater treatment plant, and the establishment of a wastewater monitoring program.

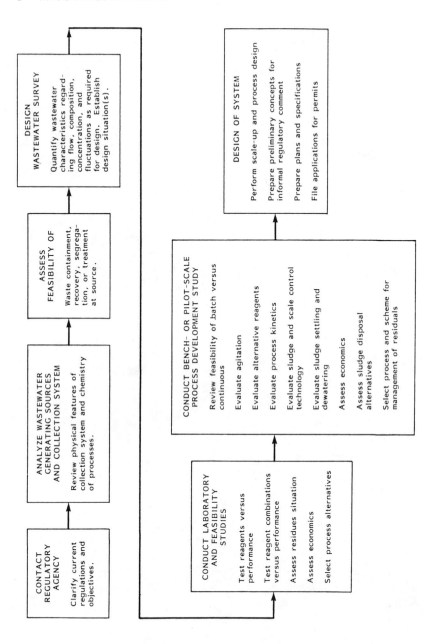

Figure 1. Procedure for the design of wastewater neutralization/precipitation systems.

Section 3 outlines the steps in establishing a design basis for a wastewater treatment plant, discusses methods of obtaining necessary information, and evaluates alternative process techniques. It provides information of particular use to the novice engineer and practicing engineers who have not been involved in the design of wastewater treatment systems.

Section 4 presents the engineering and design concepts and equations that are used for design and scale-up of neutralization and precipitation process equipment. It includes information on reagent selection, storage and handling, and construction materials. The section provides designers, plant owners and engineers, and regulatory officials with practical information needed to evaluate proposed process designs for wastewater neutralization and precipitation.

Section 5 contains construction, operating, and maintenance cost estimates--based on January 1978 dollars--for wastewater neutralization and precipitation systems. It includes cost estimates for a mixed reactor, various feed systems, and flocculation.

Section 6 presents practical information for the design and selection of sedimentation and flotation processes for the physical separation of precipitated solids from the wastewater stream. The section will prove useful to engineers and regulatory officials involved in the design, selection, and evaluation of processes for removing suspended solids from wastewater streams.

Section 7 describes methods available for treatment and disposal of the concentrated solids separated from the wastewater stream by the processes discussed in Section 6. The section provides information for plant owners, operators, and engineers who need to know the options available for sludge treatment and disposal, their advantages and disadvantages, and methods for evaluating their applicability to a particular sludge.

The appendix illustrates with examples the application of the engineering and design concepts presented in Sections 2 through 7.

2
Predesign Considerations

This section provides the plant manager and operating personnel with information that applies to the control of pH and heavy metals in wastewater. Guidance is given for implementing a water pollution abatement program, and suggestions are made for monitoring the wastewater for internal control and for conducting a survey of the plant for sources of wastewater to reduce the cost of treatment through process modification and good housekeeping.

WATER USE SURVEY

When the goals of a water pollution abatement program have been established and clearly understood, a cost-effective program must be developed to meet these goals. The first objective is to define the program in terms of water sources, uses, and disposition. This step usually involves a survey of the entire production process, which can be depicted as a material flowsheet of the plant. This flowsheet should be drawn in sufficient detail to indicate all primary discharges from each process and the type and duration of each operation. The daily or weekly discharge periods should be included. Continuous production processes should be distinguished from batch operations that have only periodic releases of wastewater. Intermittent discharges of wastewater are often important sources of pollutants and should receive as much attention as primary waste-producing operations. Figure 2 is an example of a material flow diagram that would be developed from a water use survey.

Seasonal and flow variations, including periods of peak pollution loads, should be considered. The waste characterization should identify all important parameters that yield information affecting the sampling and testing techniques to be used (i.e., high concentrations or toxic levels). The wastewater characteristics of flow, temperature, and pH also should be included.

Predesign Considerations 5

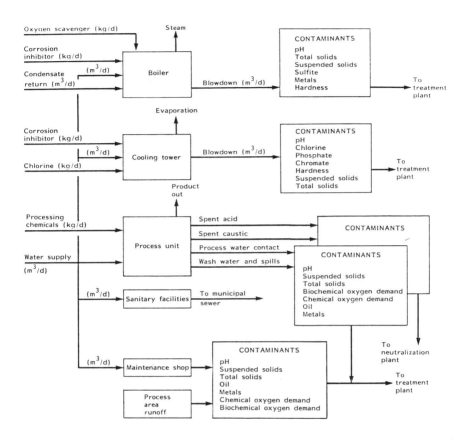

Figure 2. Example of plant material flowsheet for water use survey.

After constructing a flowsheet, it is necessary to define the amounts of raw materials, additives, products, and wastes for each operation. When the amounts of materials are known, it should be possible to establish a material balance around each production process. From the material balance, the boundaries of solid and liquid waste characteristics may be determined. A material balance for the entire plant will indicate the amount of waste generated by subtracting the amount of materials shipped from the amount purchased. This material balance acts as a check on the waste quantities determined in the preliminary survey and allows preliminary estimates of flows and parameters to be made.

Of prime importance at this point is the development of an up-to-date sewer map showing wastewater, sanitary, storm, and drain lines. The details of the map should specify pipe size, location and type of supply and drain connections to each processing unit, direction of flow, and locations of roof and floor drains, manholes, catch basins, and control points. Caution should be exercised in the use of piping diagrams because they may not have been updated as changes were made over the years.

The completed water survey will present a detailed picture of the water use and waste generation within a facility. The information from the survey can be used to design the most economical waste treatment system as well as the most effective monitoring program. Promising areas for in-process abatement efforts can also be identified. After in-plant abatement procedures have been implemented, treatment is the only remaining option. The economic break point between in-plant abatement measures and end-of-pipe treatment will differ from plant to plant, so each project engineer must formulate a balanced approach for his specific facility.

WASTEWATER MONITORING

The scope of a wastewater monitoring program can range from a survey that lasts a few days to a detailed analysis of all internal waste flows. The goal of such a program is to devise the lowest possible cost pollution control scheme. A useful guide is EPA's Handbook for Monitoring Industrial Wastewater [1].

Because wastewater flows and characteristics vary considerably with time, the initial results of a survey may be confusing. It is important that sampling and analytical procedures do not add to this confusion.

The first decision to be made in a wastewater survey concerns the amount of reliance on outside assistance. If outside specialists are engaged, an engineer experienced in plant operations should be assigned to the consultants to ensure that the rationale and intent of their analysis and recommendations are compatible with process operations. This staff engineer will present the program to the production and management staff to obtain the cooperation and assistance that is necessary for a successful project.

As the water pollution control program is being designed and implemented, it is essential that the project leader report to a management level that is high enough to guarantee that the production, analytical, laboratory, and engineering functions will cooperate fully. Otherwise, the needs of the pollution control groups may be bypassed or assigned a priority lower than the daily operating problems. Because the plant manager is responsible for meeting the requirements of any effluent permit, it is imperative that he take an active interest in the project.

During the survey, the production staff should be made aware of the need to maintain a "normal" production schedule. No waste abatement measurements should be introduced during the survey by individual actions. Water spills, waste dumps, and overflows should occur with normal frequency.

The wastewater survey and sampling program may require the full-time attention of the assigned staff. The time intervals and other circumstances peculiar to the sample procedure require constant operator attention. When automatic samplers are installed, someone must be available to maintain the apparatus and replace the sample containers. If the plant operates 24 h/d, the sample collection period should be compatible with the operation. The production staff should inform the personnel assigned to the wastewater survey of the occurrence of wastewater dumps from batch and intermittent operations. Major spills should be reported and noted to enable a proper evaluation of the results from the wastewater survey. A log should be maintained to record occurrences that may affect the sample characteristics.

When establishing a monitoring program, it is important to recognize that individual plants will have specific needs. The following three paragraphs present requirements that may or may not apply in all cases.

First, although regulatory agencies will only require monitoring of those waste streams that leave the plant site, a comprehensive monitoring program will locate inefficient and wasteful operations and lead to reduced manufacturing costs. Thus, in-plant monitoring is essential to detect changes in the process waste load in time to correct them before violations occur.

Second, in planning for monitoring needs, the same principles and process knowledge that lead to an efficient manufacturing process can be applied to the design of an optimal monitoring system. Such planning before the implementation of a monitoring program will optimize the cost effectiveness of the program while accomplishing its objectives. To minimize the analytical costs and increase the effectiveness of any monitoring program, it is essential to select the proper parameters for measurement. Although process analysis, waste surveys, analytical considerations, and choosing the proper parameters are discussed separately, each depends on the other in actual practice.

Third, the monitoring system must be designed to be compatible with projected production and waste treatment facilities. It is advisable to consider an in-plant monitoring system as a portion of a total abatement program and to be constantly alert for opportunities to minimize treatment cost while designing and implementing the monitoring program. Monitoring costs and treatment costs can be minimized by good waste management. Thus, adequate planning of the initial program will result in cost savings throughout the monitoring and treatment phases of an effective management program.

The steps involved in establishing an effluent monitoring program are illustrated in Figure 3. The wastewater monitoring program involves flow measurement, sampling, and quantitative analysis of specific contaminants.

Predesign Considerations 9

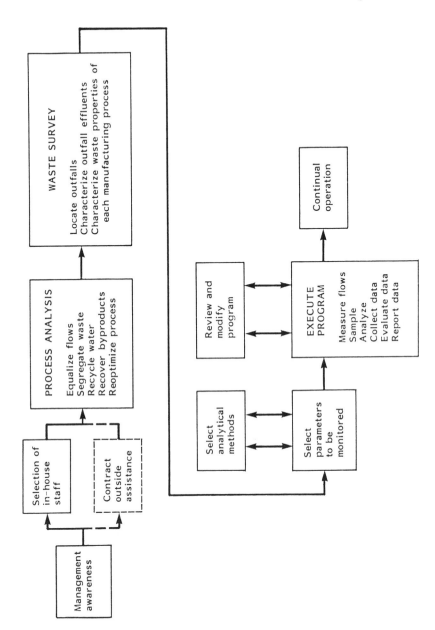

Figure 3. Steps involved in establishing an effluent monitoring program.

The procedures are similar to those described in the next subsection for characterizing the wastewater, except the monitoring program continues for the life of the plant and only EPA-approved analytical methods may be used on the streams being monitored for compliance with the National Pollution Discharge Elimination System (NPDES) permit.

Wastewater Characteristics

Characterization of the wastewater with regard to flow and concentration of pollutants is essential to the proper design of a wastewater treatment plant. Wastewater characterization involves measurement of the water flow, sampling and sample preservation, and analysis for identification and quantification of specific constituents.

Three questions must be answered when establishing a sampling program:

- What parameters should be measured?
- When should specific parameters be measured?
- Where should samples be collected?

What To Measure

The first question is answered easily: regulations usually will dictate what must be controlled. In certain cases, the parameters required by law may not be adequate for the needs of a given facility. A simple example is the measurement of complexing agents in wastewaters. Sufficient concentrations of these materials render treatment by precipitation ineffective. Thus, measurement of these species is essential to the design of a wastewater treatment facility.

When To Measure

The answer to the second question is plant specific; sampling frequency varies for continuous, batch, and intermittent processes. For plants with batch operations, the effluent may contain certain contaminants for only several minutes per day. Sampling periods must be designed to reflect such variations.

Regulatory agencies commonly require 24-h composite samples, intermittent grab samples, and, on occasion, recording of the maximal concentration for a specific parameter during a 24-h period. These requirements, however, may not be sufficient for proper waste characterization. Frequent analysis of discrete samples is required for a good approximation of wastewater characteristics and design parameters for process control.

Grab sampling is a straightforward procedure that provides an instantaneous sample of the condition of the wastewater. The problems with grab samples are illustrated in Figure 4, which shows the variation of a waste stream over five 24-h periods. The concentration of a pollutant in a grab sample can differ significantly from the average value of a parameter. Therefore, when using grab sampling, enough samples must be taken to obtain an adequate definition of trends.

To obtain a true average wastewater composition, each sample added to the composite sample must be weighted according to the flow and composition at the time of sampling. This technique, called "flow proportioning," is done differently by various samplers. Because the 24-h composite is an average value, it does not adequately represent the maximal concentration that the waste treatment system will encounter. The use of 24-h composites alone can result in unrealistic and overoptimistic designs of the treatment systems. To determine the true characteristics of a waste stream, a combination of grab and composite samples is required.

Where To Sample

To obtain a representative sample, wastewater must be collected where mixing is adequate. It may be necessary to sample various cross sections of a stream and inspect for bottom deposits or floating scum to determine where mixing is adequate. Sampling stations should be easily accessible, in an environment conducive to trouble-free operation, and located in convenient areas for plant personnel to make frequent checks on the equipment.

Once samples have been collected, they must be handled properly to obtain valid test results. Many analyses require that samples be specially preserved

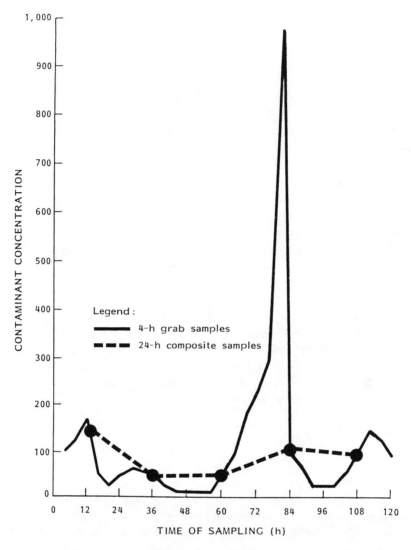

Figure 4. Wastewater contaminant concentration profiles from samples collected at 4-h and 24-h intervals.

immediately after collection. With or without preservation, samples should be analyzed as soon as possible after collection. Appropriate methods for sample preservation are described in EPA's Methods for Chemical Analysis of Water and Waste [2]. Other approved methods are published by the American Society for Testing and Materials (ASTM) and the American Public Health Association (APHA).

Well-developed techniques are available for analyzing wastewater. Approved test procedures for compliance monitoring of wastewater, in accordance with wastewater discharge permit requirements, are listed in the Code of Federal Regulations [3]. Other approved test procedures are found in APHA's Standard Methods for the Examination of Water and Wastewater [4], ASTM's Annual Book of Standards: Part 31, Water [5], and EPA's Methods for Chemical Analysis of Water and Waste [2].

Methods of flow measurement, together with their applicability to various types of problems, are summarized in Table 1.

Data Analysis

Data from the sampling and analysis program consist of daily wastewater flow records, pH records, and quantitative analysis of the pollutants present in the samples. A useful way to present these data is in a frequency distribution illustrating the pollution load variation and median. The treatment system is designed for a pollutant load that will be exceeded only a small percentage of the time. The procedure for obtaining the frequency distribution is described in this subsection and illustrated in Example 1 of the appendix.

The first step in the procedure is to tabulate the pollutant concentration and corresponding wastewater flow rate for each sample. In Step 2, the pollutant loading of each sample is calculated as the product of flow and concentration. Next, data must be normalized to the design production level. Waste loading is usually related to the production level. Each sample may have been collected at different production levels, so it is necessary to adjust the pollutant load to a common production level (e.g., kilograms of nickel per 1,000 units of product).

TABLE 1. METHODS OF FLOW MEASUREMENT AND THEIR APPLICATION TO VARIOUS TYPES OF PROBLEMS

Device or Method	Flow Range Measurement	Continuous Recording	Cost	Ease of Installation*	Accuracy of Data	Application
Mathematical formulas	Small to large	No	Low	NA	Fair	Open channels and pipe flow
Water meters	Small to large	Occasionally	Low	Fair	Excellent	Pipe flow
Bucket and stopwatch	Small	No	Low	NA	Good	Small pipes with ends or where joints can be discontinued
Pump capacity and operation	Small to large	No	Low	NA	Fair	Lines where water is being pumped
Floating objects	Small to medium	No	Low	NA	Fair	Open channels
Dyes	Small to medium	Yes	Low	NA	Fair to average	Pipe flow and open channels
Salt dilution	Small to medium	Yes	Low	NA	Fair	Pipe flow and open channels
Orifice	Small to large	Yes	Medium	Fair	Excellent	Pipe flow
Weirs	Small to large	Yes	Medium	Difficult	Good to excellent	Open channels
Flumes	Small to large	Yes	High	Difficult	Excellent	Open channels
Venturi meter	Small to large	Yes	High	Fair	Excellent	Pipe flow
Magnetic flowmeter	Small to large	Yes	High	Fair	Excellent	Pipe flow
Flow nozzles	Small to large	Yes	Medium	Fair	Excellent	Pipe flow
Pitot tube	Small to medium	Yes	Medium	Fair	Good	Pipe flow
Rotameter	Small to medium	Occasionally	Medium	Fair	Excellent	Pipe flow

*NA = not applicable.

Source: J.G. Rabosky and D.L. Koraido, "Gaging and Sampling Industrial Wastewaters," *Chemical Engineering*, 80 (1):111-120, Jan. 1973.

The normalized data are plotted on probability coordinate paper. The data are ranked in order of increasing value, that is, 1 for the lowest value and n for the highest value, where n equals the number of data points. If there are more than 20 data points, the probability of occurrence of values less than or equal to the waste load (the data point) is calculated as $(100 \times m)/(n + 1)$, where m equals the rank of the data point. If there are less than 20 data points, the lowest ranking sample is assigned a probability of $100/2n$ and each successive sample is assigned an incremental increase in probability of $100/n$. The normal distribution function or the log-normal distribution function is commonly found to represent the data.

METHODS OF WASTEWATER REDUCTION

After the wastewater survey, a plant manager should attempt to minimize pollution control costs. Housekeeping activities, which include control of water misuse, spill prevention and control, and debris cleanup, can provide measurable benefits in this area.

Flow Reduction

One of the simplest and most efficient methods for reducing the cost of a water pollution control system is to reduce the amount of wastewater being treated. Reducing the wastewater flow results in smaller and less expensive equipment, usually reduces the amount of treatment chemicals required, and improves efficiency of treatment reactions.

Flow reduction should be one of the first areas investigated when designing a water pollution control system. If flows can be reduced below certain levels, flow reduction may place a plant in a new category that is not as stringently regulated. If a plant discharges to a municipal sewerage system, it may be eligible for reclassification to some category other than "significant contributor." Flow reduction can have a number of beneficial effects on pollution control costs. For example, flow reduction has the potential to:

- minimize the size of pollution control equipment
- maximize the final concentration level required to meet permit requirements

- allow the use of less sophisticated equipment
- reduce chemical, pumping, mixing, and maintenance costs
- reduce water costs
- allow reclassification for plants discharging to a municipal system
- reduce the share of municipal treatment system cost
- improve reaction kinetics.

The first task in reducing wastewater is a critical examination of the production areas to eliminate unnecessary use of water. This program will involve educating operators as well as modifying the equipment to facilitate water saving. For example, every hose used in the plant should have a spring-loaded cutoff valve that shuts off water when released. Open hoses without nozzles can be a major source of water waste. One unattended 12.7-mm (0.5-in) hose, for example, can use 87,055 l/d (23,000 gal/d) of water. Three of these hoses can use up the total allowance of 189,250 l/d (50,000 gal/d), which will change a user's status from minor contributor to major contributor to a municipality. Disciplinary measures may be needed to discourage operators from bypassing the control device. Installing meters at each work station and requiring the station to report on water use will help reduce generation rates.

Another method of saving water is to use high-pressure spray nozzles instead of direct water streams for washing. Spray nozzles are an order of magnitude more efficient than a direct stream of water for washing or rinsing. A survey made in one poultry plant showed that total wastewater volume could be reduced 60 percent by attaching new spray nozzles.

Other ways to reduce water use include the following:

- Watch for leaking shutoff valves.
- If possible, use dry cleanup methods instead of routinely flooding areas with water. Water washdown is a necessity in most manufacturing facilities, but in many cases the water is used as a substitute for a broom.

- Reduce the water flow to each piece of equipment to the lowest level compatible with good operation, and take steps to ensure that this level is not exceeded.

After reducing the flow to rinsing or washing equipment, further improvement is usually possible by optimizing these operations. Inefficient rinsing, known as "short circuiting," occurs when only a small portion of the water used actually passes over the product. There are methods for scientifically measuring the effective flow pattern in a tank with the use of tracers. Sometimes relocation of the inlet and outlet nozzles is sufficient; in other cases, baffles are needed. In almost all cases, the method of correction is obvious after directly observing the flow patterns.

Spill Prevention and Control

Table 2 lists the causes of accidental discharges in metal finishing plants. Most spills could be prevented by careful design, use of the right equipment, and safe operating procedures. Operators and supervisory personnel should be trained in the proper methods of spill prevention and cleanup.

Areas of a plant that are vulnerable to spills should be designed with containment systems to hold the discharge in the immediate area. The drain system should be designed so that rainwater is either bled or periodically pumped to a waste treatment facility when contaminants are detected [6]. Good engineering practice also dictates that storage tanks be diked. Dikes may be simple earth structures for nontoxic materials, but concrete is preferred for containment of toxic materials. Because the potential for spillage is high in loading and unloading areas in a plant, special care should be taken to ensure that these areas are properly diked and drained. Peripheral trenching covered with grating is also useful for collecting and disposing of spills that occur with tank truck and tank car operations.

The plant monitoring system should be designed to alert plant operating personnel when a spill occurs to enable them to take immediate corrective action.

TABLE 2. COMMON WASTE DISCHARGES CAUSED BY ACCIDENTS IN METAL FINISHING PLANTS

Source	Method of detection	Correction or prevention

Process tank overflow:
  Unattended water additions
  Leak of cooling water into solution from heat exchanger of cooling coil
	High-level alarms in floor collection systems to signal unusual discharges
	
	Integrated floor spill treatment
		Provide proper floor construction for floor spill segregation and containment (curbs, trenches, pits)
		
		Provide treatment facilities for collected floor spill
		
		Integrated floor spill treatment system
		
		Use of spring-loaded valves for water additions
		
		Provide automatic level controls for water additions

Process solution leakage:
  Tank rupture or leakage
  Pump, hose, pipe rupture or leakage, filtration, heat exchanger, etc.
  Accidental opening of wrong valve
	High-level alarms in floor collection systems to signal unusual discharges
	
	Integrated floor spill treatment
		Provide proper floor construction for floor spill segregation and containment (curbs, trenches, pits)
		
		Provide treatment facilities for collected floor spill
		
		Integrated floor spill treatment system

Normal drippage from workpiece during transfer between process tasks	Inspection	Provide drainage pans between process tanks so that drippage returns to the tanks Collect floor spillage Integrated floor spill treatment
Process solution entering cooling water (heat exchanger leak)	Conductivity cell and bridge to actuate an alarm Use of the cooling water as rinse water in a process line where the contamination will be immediately evident	
Process solution entering steam condensate (heat exchanger or heating coil leak)	Conductivity cell and bridge to actuate an alarm	Use conductivity controller to switch contaminated condensate to a waste collection and treatment system
Spillage of chemicals when making additions to process tanks or spillage in the chemical storage area	Make the solution maintenance man responsible for chemical additions	Careful handling and segregation of chemical stores Segregation and collection of all floor spillage Integrated floor spill treatment

SOURCE: Lancy Laboratories, The Capabilities and Costs of Technology Associated With the Achievement of the Requirements and Goals of the Federal Water Pollution Control Act, as Amended, for the Metal Finishing Industry, National Commission on Water Quality, 1975.

Countercurrent Operations

Further reduction of wastewater flow can involve process modification. An example of a process modification is the concept of countercurrent operation, which is used by plants with extensive rinsing operations.

Figure 5 illustrates three rinsing operations, all designed to remove the residual acid on the surface of the workpiece. In the first application (Figure 5a), the piece is dipped in one tank to which water is flowing continuously. The acid on the surface of the workpiece is being diluted to the required level. Assuming complete mixing, the amount of water theoretically required to remove 4.5 kg (10 lb) of acid on a 9-m^2 (100-ft^2) surface with a film drag of 11 g/cm (1 oz/in) is about 4,542 l (1,200 gal) of water [7].

In Figure 5b, fresh makeup water is used in a final rinse before the article moves out of the rinsing section. The slightly contaminated water is used again in a second stage to clean articles entering the rinsing section. Assuming well-mixed tanks and a limit of 0.1 percent acid on the first stage rinse, the double counterflow rinse requires 454 l (120 gal) of water. If the number of stages is increased to three (Figure 5c), further reductions in water use are realized. The theoretical reduction of water by countercurrent multistage operation is shown in Figure 6. The actual flow reduction obtained is a function of the drag-out and the type of contact occurring in the tanks. Large amounts of short circuiting, for example, make such reductions impossible. With efficient rinsing, it is reasonable to expect major reductions in water use. Figure 6 illustrates that the largest reductions are made by adding the first few stages.

There are many configurations other than direct staging that can be used to approach the countercurrent system. For instance, in the dip and spray technique, the workpiece is dipped into a rinse tank to remove the bulk of the contaminants and is sprayed with clean water as it is removed. The water dripping from the piece into the tank provides the dilution necessary to control the amount of contaminants reaching the freshwater spray.

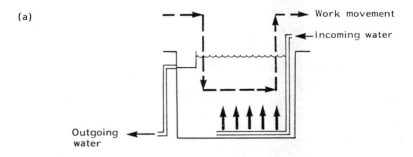

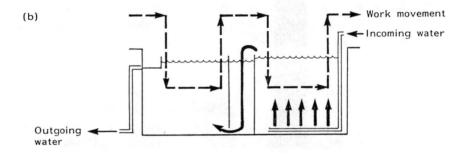

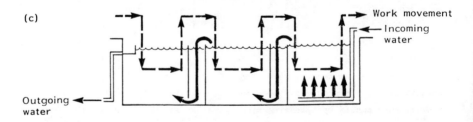

Figure 5. Counterflow rinsing operations: (a) single, (b) double, and (c) triple.

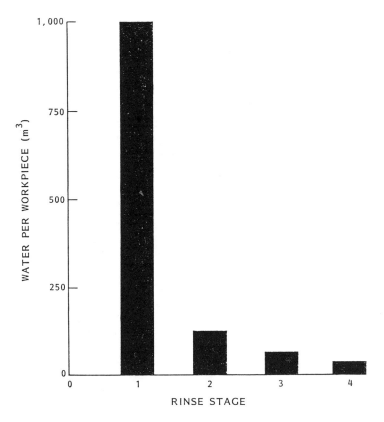

Figure 6. Effects of an increasing number of rinse stages on water use.

Another factor that must be considered when installing multistage rinse systems is the need for sufficient agitation. If water is being cascaded in enormous quantities over a workpiece, the high flow usually provides agitation. As the amount of water is reduced, however, a point is reached where the flow does not provide sufficient agitation. In this case, either careful baffling of the tanks or additional agitation (mechanical or air) is required.

Flow Equalization

Another way to minimize cost in a wastewater treatment system is to equalize the flow. Even after maximal application of in-plant controls, most plants have a wide variation in the flow of wastewater. Flow variations are particularly high if a manufacturing process includes many batch operations. Continuous processes have fewer variations, yet flow variation is still a significant factor in the design of pollution control equipment. Flow equalization can reduce wide swings in pollutant concentrations, allowing the wastewater treatment system time to respond.

Flow equalization balances wastewater characteristics and composition to a certain degree, but this equalization is limited because the only agitation provided is that caused by inflow and outflow. Although continuous systems that would completely equalize contaminant concentrations are not practical, the averaging effect that results from mixing the various waste streams may determine the success or failure of the treatment system.

Wastewater flow can be equalized in a number of ways. Some equalization occurs in the sewer lines. Pollution control equipment with a given surge capacity also will dampen out some variations. In terms of wastewater pollutants, however, large variations may still occur if concentrations entering the flow equalization system are excessive.

Example 2 in the appendix describes the procedure for sizing the storage capacity needed to maintain a fixed flow to the wastewater treatment system.

Other equalization concepts designed to minimize the variability of the waste stream to be processed are batch processing, continuous processing from batch storage and equalization tanks, side-stream equalization, and flow-through equalization, as illustrated in Figure 7.

Batch Processing

Batch processing involves storage and batch processing of the wastewater in the same vessel. The system is designed with multiple reactor vessels; when one vessel is full, the flow is switched to another vessel while the

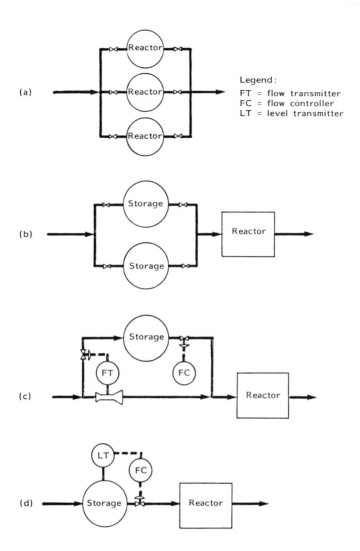

Figure 7. Alternative concepts for wastewater equalization: (a) batch reactor system, (b) batch equalization/continuous processing, (c) side-stream equalization, and (d) flow-through equalization.

wastewater in the first vessel is treated. Vessel storage capacity must be sufficient to handle the maximal volume of wastewater that could be generated during the time it takes to treat the wastewater and empty it from the vessel.

Continuous Processing From Batch Storage

Continuous processing usually involves two storage tanks operating on a fill-and-draw cycle. While one tank is being filled, the other tank is being discharged to a waste treatment facility. The waste treatment facility receives a waste stream with uniform characteristics while processing the contents of an individual tank. The stream characteristics may vary for each tank, however.

Side-Stream Equalization

The side-stream equalization method involves the collection and storage of flow surges or concentrated wastes that result from scheduled dumping operations in the production plant. The flow surge or concentrated waste is sensed in the transmission line and diverted to a storage vessel for subsequent processing during slack periods. This procedure could be used to store an alkali waste in a plant with a waste stream that is normally acidic. The collected alkali could be dispensed into the normally acid waste stream to reduce consumption of neutralizing reagents in the wastewater neutralization facility.

Flow-Through Equalization

The system of flow-through equalization reduces but does not eliminate variations in wastewater flow and concentration. The objective of this system is to attenuate sudden changes in the wastewater characteristics that would otherwise adversely affect process control. Flow-through equalization is accomplished by placing a resistance, such as a valve or a weir, in the flow path.

A simple flow resistance can be provided by a proportional weir, the flow from which can be characterized by the linear equation:

$$F = kh \qquad (1)$$

where F = volumetric flow rate through the weir
   k = a discharge coefficient
   h = level of liquid behind the weir above a reference point

A resistance that is characterized by a linear equation is a linear resistance. Under steady operating conditions, the flow rate of the water behind the weir is equal to the flow rate through the weir. If the flow rate behind the weir changes, however, the flow through the weir also changes, though not at the same rate or in the same characteristic manner. The changes in flow rate assume that no buildup of solids occurs behind the weir.

The weir effectively modified the rate of change in the flow rate downstream of the weir. The manner in which the flow rate through the weir changes can be determined by transient flow analysis techniques, a discussion of which is beyond the scope of this manual. Such an analysis would yield the following equation, which describes the flow through a linear resistance following a sudden (step) change in the flow to the weir:

$$F_w = F(1 - \exp^{-kt/A}) \qquad (2)$$

where F_w = flow through the weir
   F = magnitude of the step change
   A = cross-sectional area of water behind the weir
   t = time since the step change occurred
   k = constant

Figure 8 illustrates Equation 2.

Underestimation of the variability of the waste is the main reason why pollution control systems fail to meet expectations. The designer must know both the peak values and the rate of change of flows that will affect his equipment. Flow equalization can minimize the risk of overloading the system.

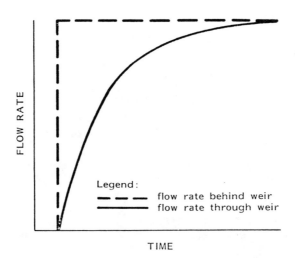

Figure 8. Effect of a linear resistance (proportional weir) on flow rate.

Further details on the design and application of equalization systems are available in articles by Di Toro [8], Novotny and Englunde [9], and Novotny and Stein [10].

Stream Segregation

In many manufacturing plants a large portion of the wastewater is cooling water. If the cooling water does not come into contact with any pollutants, EPA classifies it as noncontact water. A noncontact water stream is not part of the process wastewater load and, unlike contact water, does not require stringent monitoring and treatment.

Typical causes of contamination of cooling water include:

- oil leaks into cooling water pumps
- direct contact condensers

- process leaks into cooling water
- mixing of process waste with cooling water because of inadequate sewage system.

If a significant portion of once-through cooling water becomes intermingled with a waste stream, it is advisable to look for methods to prevent this contamination.

The intent of stream segregation is to tailor waste treatment methods to the specific problems of individual effluent streams. The advantages of this approach include:

- Pollutants are treated at maximal concentration, which increases treatment efficiency.
- Interactions between waste streams (i.e., complexation of metals in one stream by chelating agents present in another) that hinder treatment are minimized.
- Recovery of stream components, for reuse or resale, is enhanced.

The disadvantages of such a system are added piping and repiping costs and the capital costs of individual waste treatment units, including the additional operating and maintenance costs.

Stream segregation is being practiced in many industries. It would not be logical to combine wastes before treatment if all the pollutants cannot be destroyed in the same process. Stream segregation is also practiced to avoid combining incompatible wastes.

Before evaluating the usefulness of stream segregation, the chemical composition of each stream must be known. The recovery of a stream constituent should be assessed first. If one chemical accounts for the major portion of the waste load from a stream, recovery may be practical and should be investigated.

Although separate treatment of waste streams within a plant is usually impractical, segregation is feasible for many streams and often results in more efficient and effective wastewater treatment. As a rule, the efficiency of metal removal is directly proportional to the initial concentration. Mixing a concentrated waste stream with a dilute waste stream (regardless of the volume of each stream) causes a net decrease in treatment efficiency. This effect is particularly pronounced when mixing streams that contain chelating agents. Therefore, streams containing metals should not be combined with streams containing compounds that may form strong complexes with metals. Some of these compounds are ethylenediaminetetraacetic acid (EDTA), citric acid, nitrilotriacetic acid (NTA), tartrates, thiourea, gluconic acid, ammonia, cyanides, and polyphosphates.

Safety is another reason for stream segregation. Effluents containing cyanides or sulfides can be dangerous because, if acidified, they form gases that are toxic.

The cost of stream segregation detracts from its desirability. These costs are site specific and must be calculated individually.

Water Reuse

The entire manufacturing sequence, rather than individual processes, should be examined to determine if water used in one application is suitable for reuse in other areas that require less purity. Determining the amount of water to be reused in a plant requires knowledge of the water quality requirements for each process using water. Precise figures often are not known, but engineers or operators familiar with a process can usually make adequate estimates. Careful experimentation may also be required. This inventory of water quality requirements is then matched with the characteristics of wastewater being generated.

In many plants, the addition of surge capacity will be necessary because production areas may not be operating in a matched sequence. In some cases, product quality will be affected when contaminated water is used. Testing of each application should proceed slowly over a period of time to avoid the

LABORATORY EVALUATION

Methods

The most widely used test to determine optimal conditions for precipitation is the jar test, which attempts to simulate the full-scale precipitation-coagulation process. Since its introduction in 1918, the jar test has remained the most common control test in the laboratory. Test apparatus consists of a series of sample containers (usually six), the contents of which can be stirred by individual mechanically operated stirrers. Wastewater to be treated is placed in the containers, and treatment chemicals are added while the contents are being stirred. The range of conditions (e.g., precipitating reagent concentration, coagulant dosages, and pH) is selected to bracket the anticipated optima. After a 1- to 3-min period of rapid stirring to ensure complete mixing, the stirring rate is decreased and coagulation is allowed to continue for a variable period--5 to 15 min or more, depending on the simulation. The stirring is stopped, the flocs are allowed to settle for a selected time, and the supernatant is analyzed for the desired parameters--typically, heavy metal concentrations, turbidity of suspended solids, pH, and residual coagulant.

If desired, a number of supernatant samples may be taken at intervals during the settling period to permit construction of a set of settling curves, which provide more information on the settling characteristics of floc than a single sample taken after a fixed settling period. A dynamic settling test also may be used in which the paddles are operated at 2 to 5 r/min during the settling period. The dynamic settling test more closely represents settling conditions in a large horizontal basin with continuous flow. It should be noted, however, that simple jar tests cannot simulate the conditions in solids contact reactors and may indicate somewhat higher coagulant dosages than are actually necessary when using these units for coagulation.

Reagent Selection

Although neutralization and precipitation are separate and distinct unit processes, it is usually advantageous to combine these processes into a single-unit operation. (An exception would be acid and alkaline process waters that can be conveniently combined before treatment.) The choice of appropriate reagents for single-unit operation is somewhat restricted; commonly used reagents include calcium oxide, calcium hydroxide, calcium carbonate, sodium hydroxide, sodium carbonate, sodium hydrogen carbonate, and various soluble and insoluble sulfides. Additionally, there is a choice of coagulants for each neutralization/precipitation system. This subsection provides a rationale for making such a selection. Specific reagents, however, will be discussed in Section 3.

The choice of reagents is based on desired effluent limitations and economic restrictions. It is possible that two reagents may satisfy legal discharge requirements. Where either choice is appropriate, the decision is made on the basis of reagent cost, equipment cost and reliability, and ease of sludge handling and disposal.

The reaction conditions (solution pH, temperature, and coagulant dosage) during precipitation determine the properties of the resulting sludge. At present, however, theories of precipitation and coagulation are not sufficient to predict the precipitate (sludge) properties as a function of reaction conditions without experimentation.

Figure 9 shows characteristic settling curves that may be obtained. Curve A indicates a coagulation that produced a uniformly fine floc so small that, at the end of 1 to 2 min of settling, the supernatant had a turbidity equal to that of the starting water. Settling was slow and the final turbidity was not satisfactory. Curve B represents the most common type of settling rate obtained. During the first 5 min, the settling rate was practically a straight line on a semilog plot. Settling was rapid and clarification was satisfactory. The coagulation represented by Curve C shows that a mixture of large, rapid-settling floc and small, slow-settling particles was obtained. Settling was rapid for the first 2 min, but little clarification occurred after that. High residual turbidity also may have resulted from incomplete coagulation. Curve D represents the ultimate in coagulation. Practically all of the floc particles were so large and dense that 97 percent settled within 3 min. Sedimentation was essentially complete within that time because only 0.5 percent additional floc settled in the next 27 min. Final clarity of the supernatant was entirely satisfactory.

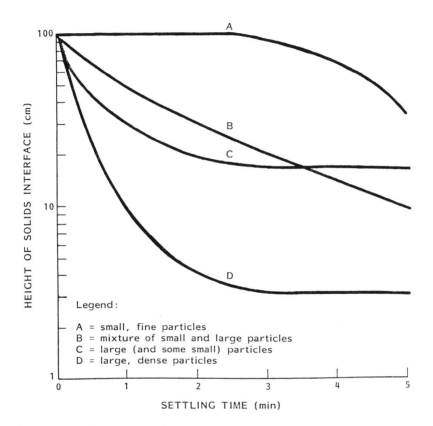

Figure 9. Examples of settling curves.

Measurement of turbidity provides the most rapid indication of the degree of solids removal obtained where the character of the solids does not vary widely, and concentration generally correlates well with measured turbidity. The recommended procedure for turbidity measurement by light scattering is given in <u>Standard Methods for the Examination of Water and Wastewater</u> [4]; however, other methods--varying from simple visual evaluation to measurement of light transmitted on a turbidimeter or, if not available, a laboratory spectrophotometer--can be used for comparison. Measurement of residual suspended solids is the only procedure that gives the actual weight concentration of solids remaining, but the procedure is too slow for process control use.

After determining an approximate optimal coagulant concentration, it may be desirable to repeat the jar test using that optimum while varying pH. Experience in coagulating a given wastewater provides the best guide of methods for controlling the process.

SUMMARY

From the preceding discussion, several principles of water management can be summarized as follows:

- Means must exist to monitor and record flow rates through the system. From this information, it is possible to analyze where problems are occurring, where water use may be reduced, and where things are going according to plan [11].

- Good engineering practice must be applied to water management. The same knowledge, physical laws, and problem-solving approaches that are used to solve process problems can be used to reduce the amount of aqueous discharges from a process or a plant. In a broader sense, however, good engineering practice considers an effluent problem within the context of process optimization [11].

The most commonly demonstrated techniques for lowering the total cost of water pollution control have been discussed in the context of practices to minimize the amount of pollutants that must be treated. It is an economic necessity to investigate the use of in-plant control techniques before installing any treatment system.

For some plants, in-plant controls will be an effective way to minimize pollution control costs and to meet requirements that otherwise would be impossible to reach. Most plants, however, eventually will reach a point where the cost of further flow and pollutant reduction by in-plant control becomes too expensive. The remaining sections of this manual, therefore, are concerned with the design and treatment of an optimized waste stream.

3
Process Development

Designing a wastewater treatment plant involves the economical application of design fundamentals to a practical situation. Figure 10 illustrates the steps in process development and evaluation. This section outlines the establishment of a design basis, including a discussion of information needs and the methodology for obtaining the information, and an evaluation of alternative process techniques. Sample problems are included in the appendix to illustrate the use of the information presented.

The reader is cautioned that there is no universal stepwise procedure that can be applied to the design of wastewater neutralization and precipitation systems. Each process plant has unique characteristics that must be considered when designing a plant for a particular situation. Information needed for sizing the neutralization reactor includes wastewater flow rates, reagent type and quantity, operating conditions, reaction kinetics, and method of operation. Procedures for determining wastewater flow rates and pollutant loading are fundamental and will not be discussed in this manual. Before studying the reaction kinetics, it is necessary to select the reagent that will be used for neutralization.

SELECTION OF REAGENT

Factors to consider in selecting a suitable reagent for wastewater neutralization processes include speed of reaction, buffering qualities, product solubility, reagent cost and availability, ease of reagent handling, and costs associated with reagent handling. The properties, advantages, and disadvantages of some common neutralizing agents are presented in Table 3. Brief descriptions of some common neutralizing agents follow.

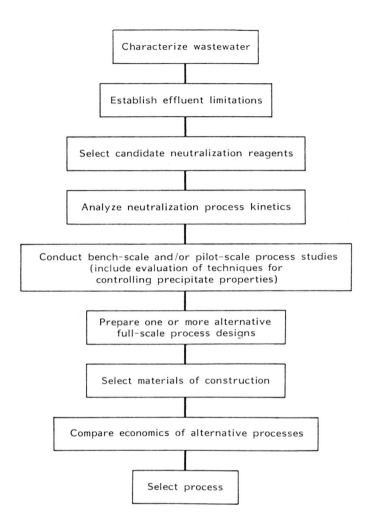

Figure 10. Steps in process development and evaluation.

TABLE 3. COMMON REAGENTS FOR NEUTRALIZATION AND PRECIPITATION

Chemical Name, Formula, and Trade Name	Molecular Weight	Equivalent Weight	Common Forms and Commercial Strength	Bulk Density (kg/m^3)	Solubility (g/100 g water at °C)	Typical Use	Advantages	Disadvantages
Ammonia (NH_3)	17.0	17.0	Gas		53.1 at 20	Acid neutralization	High solubility, Rapid reaction	Nutrient
Calcium carbonate ($CaCO_3$), high calcium limestone	100.1	50.05	White or gray: powder about 95% granules $CaCO_3$	1,602 to 1,842	0.0014 at 25	Acid neutralization	Relatively inexpensive	Contains impurities, slow reacting, difficult to handle
Calcium hydroxide [$Ca(OH)_2$], high calcium hydrated lime	74.1	37.05	White powder: 72 to 74% as CaO	400–560	0.18 at 0 0.15 at 30 0.07 at 100	Acid neutralization	Relatively inexpensive	Contains impurities, slow reacting, difficult to handle
Calcium oxide (CaO), high calcium quicklime	56.1	28.05	White: lump, pebble, granulate or pulverized (93 to 98% CaO)	881–961	Converted to $Ca(OH)$	Acid neutralization	Relatively inexpensive	Requires storage in dry atmosphere
Carbon dioxide (CO_2)	54	27	Clear, colorless gas becomes liquefied under pressure			Alkali neutralization	May be cheaply available from flue gases	
Dolomitic hydrated lime [$Ca(OH)_2 \cdot MgO$]	114.4	33.9	White powder: 46 to 48% CaO and 33 to 34% MgO	400–560	0.18 at 0 0.15 at 30	Acid neutralization	Relatively inexpensive	Less reactive than high calcium lime
Dolomitic quicklime (CaO·MgO)	96.4	24.9	White (light gray, tan): lump, pebble (55 to 57.5% CaO), or pulverized (37.6 to 40.8% MgO)	881–961	Converted to $Ca(OH)_2$ and MgO	Acid neutralization	Relatively inexpensive	Less reactive than high calcium lime
Sodium carbonate (Na_2CO), soda ash	106.0	53.0	Anhydrous: white powder (58% Na_2O)	560 to 1,041	7.1 at 0 18 at 20 27.6 at 30	Acid neutralization	Highly reactive, Higher soluble	Higher cost than calcium reagents

(continued)

TABLE 3. COMMON REAGENTS FOR NEUTRALIZATION AND PRECIPITATION (concluded)

Chemical Name, Formula, and Trade Name	Molecular Weight	Equivalent Weight	Common Forms and Commercial Strength	Bulk Density (kg/m^3)	Solubility (g/100 g water at °C)	Typical Use	Advantages	Disadvantages
Sodium hydroxide (NaOH), caustic soda	40.0	40	Anhydrous: white solid or flakes (76% Na$_2$O), liquid (50% NaOH)		42 at 0 347 at 100	Acid neutralization	Highly reactive, convenient handling	Higher cost than calcium reagents, must be heated in storage
Sodium pyrosulfite (Na$_2$S$_2$O$_5$), sodium metabisulfite	190.1	95.05	Anhydrous: powder (62% SO$_2$), 20 to 25% SO$_2$	1,185 to 1,362	54 at 20 81.7 at 100	Chromate removal processes		
Sodium sulfide (Na$_2$S·9H$_2$O)	240.2	120.1	Reddish or yellow solid, or flakes (60 to 62% Na$_2$S)		125 at 25	Acid neutralization Heavy metal precipitation	Strong alkali	May liberate toxic H$_2$S
Sulfur dioxide (SO$_2$)	64.1		Colorless gas: 100% SO$_2$		16.2 at 25 7.8 at 30	Chromate removal processes		
Sulfuric acid (H$_2$SO$_4$)	98.1	49.05	Liquid: 60°Be–77.7% H$_2$SO$_4$ 60°Be–93.2% H$_2$SO$_4$	1,704 1,834	Complete	Alkali neutralization	Highly reactive, inexpensive	Forms calcium sulfate sludge with calcium alkali
Hydrochloric acid (HCl), muriatic acid	36.5	36.5	Liquid: 20°Be 22°Be 23°Be	1,157 1,177 1,186	Complete	Alkali neutralization	Highly reactive	More expensive than sulfuric acid

Source: Adapted from W.A. Parsons, *Chemical Treatment of Sewage and Industrial Wastes*, National Lime Association, Washington DC, 1965.

Ammonia

Ammonia is a highly soluble gas that reacts rapidly and forms soluble reaction products. Because the discharge of ammonia is often restricted by nutrient considerations, neutralization usually is restricted to operations that include recovery of the byproduct ammonia salt. Ammonia can be stored and fed as a gas, liquid, or aqueous solution.

Carbon Dioxide

Carbon dioxide can be obtained in bulk as a liquid but should be vaporized for use. Stack gases from fossil-fuel-fired furnaces contain carbon dioxide and can be used to neutralize alkaline wastes. Because a solution of carbon dioxide in water produces carbonic acid, diffuser pipes for the carbon dioxide should be made of a noncorrodible material.

Lime Materials

Lime is a term used to designate calcined or burned limestone (quicklime or CaO) and its hydrated derivative [hydrated lime or $Ca(OH)_2$]. The two basic types of limestone used are high calcium and dolomitic. High calcium limestones consist chiefly of calcium carbonate with a small amount of magnesium carbonate. Dolomitic limestones contain nearly equal molar qualities of calcium and magnesium carbonate. Typical impurities present in the limestone in amounts of less than 5 percent include silica, iron, and alumina.

The reactivity of limestones differs with physical characteristics (e.g., size and shape) and chemical composition. As a rule, high calcium limestones are more reactive than dolomitic limestones; however, pronounced physical characteristics may produce exceptions to the rule. Theoretically, dolomitic limestone has greater basicity, but actual or available basicity depends on the conditions of application. Pulverized limestone is a stable noncorrosive product that is amenable to dry feeding. It is available in bulk or in 36-kg (80-lb) bags. Various mechanical conveyors are suitable for unloading the bulk material.

Quicklime

Quicklime has an affinity for carbon dioxide and water. Under improper handling or storage conditions, quicklime will air-slake, that is, absorb moisture and carbon dioxide from the atmosphere, causing physical swelling and a marked loss of chemical activity. Consequently, quicklime must be stored in moisture-proof areas that are free from carbon dioxide. Because of this perishability, quicklime is usually consumed within a few weeks after manufacture.

Quicklime is available in various forms, ranging from 20-cm (8-in) lumps to pulverized, and is supplied in bulk or in 36-kg (80-lb) bags. Dust from pulverized quicklime can irritate eyes and skin. Although it can be fed dry, for optimal efficiency it is slaked (hydrated) and slurried before use under conditions that will yield maximal reactivity. Improper slaking will adversely affect reactivity. Slaking usually is carried out at temperatures of 82° to 99° C (180° to 210° F). The slaking reaction may reach completion in 10 min with highly reactive limes or in more than 30 min for limes of lower reactivity [12].

Following slaking, the lime putty usually is slurried with water to a concentration of 10 to 35 percent (based on dry solids) for feeding purposes. Because the slurry is subject to deterioration from carbonation during storage, it is customary to use it soon after it is made.

The applicability of lime to specific situations may be expected to vary significantly from supplier to supplier. Testing under actual or simulated process conditions is the only sound basis for determination of relative applicability; empirical basicity tests normally are of value only when the application is analogous to the test.

Hydrated Lime

Hydrated lime is suitable for dry feeding or for slurrying. Dust from hydrated lime, a fine powder, can cause eye and skin irritation. The storage characteristics of dry hydrated lime are superior to quicklime, but, as with any strong alkali, carbonation can cause deterioration. Hydrated lime is supplied in bulk or in 23-kg (50-lb) bags. Bulk unloading is usually accomplished by pneumatic conveyor.

Sodium Carbonate

Sodium carbonate (soda ash) is a highly reactive soluble alkali that is marketed most often as an anhydrous powder. Wet crystal bulk storage facilitates solution feeding. In dry form it is also easily fed from hoppers. Positive provision for dissolution is desirable for dry feed applications. Suitable materials for handling the compound or its solutions include plastic, iron, rubber, and steel. Shipment is made in bags, barrels, or in bulk with transfer usually performed by pneumatic conveyor.

Sodium Hydroxide

Sodium hydroxide (caustic soda) is a highly reactive alkali that is marketed in solid or solution form. The solution form is the most convenient for handling because burn hazards to personnel are minimized. The solid form is hygroscopic, and both the solid and the solution are subject to deterioration from carbonation during prolonged storage. Solid and liquid sodium hydroxide is supplied in drums, but only the liquid form is available in bulk (tank car or truck). Heated tanks should be used for storage of 50-percent solution in situations where the ambient temperature is likely to fall below 12° C (54° F).

Sodium Metabisulfite

Sodium metabisulfite (pyrosulfite), a potent reducing agent, is available in both powder and solution form. It is usually fed in solution form and can be handled safely in plastic, glass, lead, rubber, and ceramic containers.

Sodium Sulfide

Sodium sulfide, which is flammable and strongly alkaline, reacts with metal ions to form insoluble metal sulfides. It is a skin irritant and liberates toxic hydrogen sulfide on skin contact or when made acidic. Sodium hydrosulfide is an alternate sulfide reagent that is available in flake and liquid forms.

Sulfur Dioxide

Sulfur dioxide is a gaseous reducing agent that is generally marketed in cylinders weighing from 45 to 907 kg (100 to 2,000 lb). It is usually stored in its shipping container and fed by specially manufactured feeding equipment. Aqueous solutions of sulfur dioxide are acidic and corrosive. Suitable materials for handling are plastics, glass, lead, nickel, and rubber.

Sulfuric Acid

Sulfuric acid is a highly reactive acid that is supplied in liquid form, usually in concentrations of 98 percent. The concentrated acid is strongly hygroscopic and presents a burn hazard to personnel. Dilute solutions are highly corrosive to iron and steel, whereas concentrated solutes (>93 percent) are not corrosive. A maximal freezing point of 8° C (47° F) is exhibited at a concentration of 85 percent. Therefore, solutions should be protected from freezing during storage and transport. Sulfuric acid is shipped in carboys, barrels, tank cars, or trucks.

DETERMINATION OF REAGENT QUANTITY

Following the selection of candidate reagents, the quantity of reagent required to neutralize the waste stream must be determined. Reagent quantity is calculated by developing titration curves for representative wastewater samples using the candidate reagents. In some cases (particularly lime) the reagents from different suppliers should be tested separately. Titrations should be performed on a number of wastewater samples that represent the range of conditions likely to be encountered in the plant. The cumulative amount of reagent added and the pH of the sample after each addition are recorded and graphically plotted as shown in Figure 11. The data should be converted to represent the amount of reagent--on a 100-percent basis--required to bring the sample volume of wastewater to the desired pH. The amount of reagent at the concentration that will be used in the actual plant can then be readily calculated.

The preparation of reaction rate curves for each candidate reagent is the next step in the experimental procedure. It is desirable that these experiments be performed in an apparatus dimensionally similar to the standard

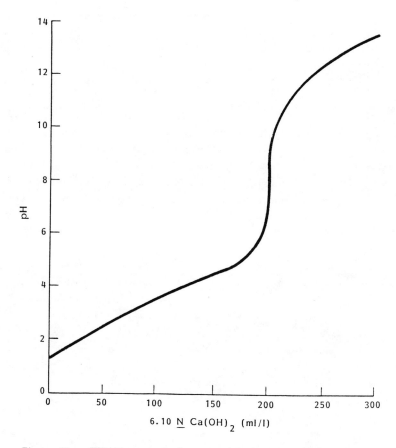

Figure 11. Titration curve for neutralization of process waste containing strong acids with 6.10 $\underline{N}$ Ca(OH)$_2$.

reactor illustrated in Figure 12 (including baffles, agitator, etc.) if scale-up is to be performed. A transparent experimental reactor is desirable.

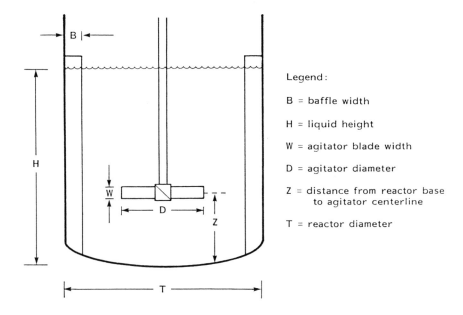

Legend:

B = baffle width

H = liquid height

W = agitator blade width

D = agitator diameter

Z = distance from reactor base to agitator centerline

T = reactor diameter

Note.--The ratios of dimensions relative to standard geometry are given in Table 4 in the next section.

Figure 12. Standard agitated reactor.

The initial experiments are designed to determine the relationship between pH and time for selected reagent dosages at the mean process temperature and with an agitation intensity sufficient to maintain suspension of solids above the vessel floor. The amount of reagent required to reach the desired pH is added all at once to the wastewater mixture in the reactor. The mixture is stirred continually while its pH is recorded continually as a function of time. Other observations that may be made include reagent dosage, temperature, temperature rise, agitator speed, density, viscosity, color, turbidity, conductivity, and residual target substance. If applicable, sludge volume and characteristics would be observed, as well as any tendency to form scale deposits. Excessive temperature rise may require dilution or other temperature control measures, such as indirect cooling.

After determining the initial rate curve, a family of curves (Figure 13) can be developed to identify the curve that will give the best design to effect the required pH change. Assuming pH guidelines of 6.5 to 9, a rate curve that breaks at a pH of 6.5 and becomes asymptotic to a line drawn through a pH of 9 would be desirable when neutralizing an acid waste. (For different pH guidelines, the appropriate values would be substituted.) A rate curve that passes through pH values of 6.5 and 9 but continues on until a pH of 11 or 12 would require the ability to stop the reaction at 9 or adjust the pH back from 11 or 12. (This technique has been employed successfully.)

Four or five different test runs might be required to obtain the rate curve that will best fit limitations on pH. Each test run should be conducted as described for the initial run, except that different stoichiometric quantities of the reagent will be added to the samples. Usually, a variance of 1 to 5 percent by weight of the amount of reagent should be enough to define the proper curve.

The dosage of the first run is 100 percent. If the stoichiometric quantity of reagent is decreased or increased by 1 to 5 percent for each subsequent run, the other runs are labeled accordingly.

In Figure 13, Curve B best meets the preceding criteria. Curves D and E can be eliminated because they do not bring the waste effluent sample into the desired pH range. Curve A exhibits the fastest reaction rate but would require a design that could stop the reaction at a pH of 9 or provide for a means of lowering the pH from 13 to the range of 6.5 to 9. Curve C would provide an acceptable design in terms of the desired pH range, but Curve B also is within the design limits. Because Curve B exhibits a faster reaction rate and shorter reactor residence time than Curve C, a smaller reactor is

Process Development 45

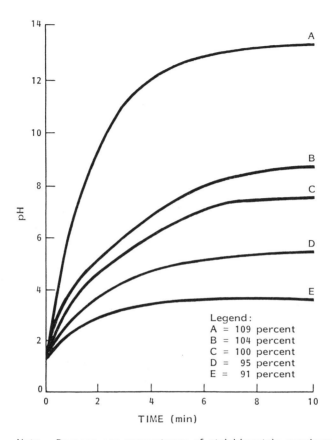

Note.—Dosages are percentages of stoichiometric requirement.

Figure 13. Sample reaction rate curves for addition of high calcium lime slurry to process waste containing strong acids.

possible by using the reagent dosage represented by Curve B. Therefore, the reactor design would be based on Curve B.

From the titration and reaction rate curves, a curve of residual reagent as a function of time can be developed. The amount of reagent required to reach the desired pH is determined from the titration curve (Figure 11). The reaction rate curve is based on adding some percentage of this amount. Therefore, the total dosage of reagent can be calculated. From Figure 11, the amount of reagent [i.e., ratio of 6.1 $\underline{N}$ Ca(OH)$_2$ to wastewater] required to reach a pH of 7 is 200 ml/l (26 oz/gal). The reaction rate curve is based on adding 104 percent of the theoretically required dosage (1.04 X 200), or 208 ml/l (27 oz/gal). At any time, t, the wastewater has a particular pH. The titration curve shows the amount of reagent that has reacted with the wastewater at any pH. By subtracting this amount of reagent from the amount of reagent originally added, the residual reagent is determined. Figure 13 shows that for Curve B, at 0.5 min, the pH of the wastewater is 3.45. From the titration curve, at a pH of 3.45, 100 ml/l (13 oz/gal) of a 6.10 $\underline{N}$ Ca(OH)$_2$ solution has reacted with the wastewater sample. The residual reactant at 0.5 min is therefore 208 ml minus 100 ml, which equals 108 ml/l (14 oz/gal) or 24.38 g/l (3.12 oz/gal).

This calculation would be repeated at various time intervals along the entire pH-versus-time curve (Curve B of Figure 13) to yield a ratio of residual reagent versus time. It is important to gather accurate data in the region of the equivalence point. In many cases, it is necessary to work directly from titration data to show the minute changes in reagent as the pH of the sample changes.

It is useful to plot the relationship between residual reagent and pH. Because all calculations for reactor design involve residual reagent, these values will have to be related to a pH value. The graph usually can be drawn on semilog paper because values of residual reagent span a large domain.

After translating the pH-versus-time curve to a residual reagent relationship, a kinetic rate equation must be developed. The several mathematical and graphical techniques that may be employed to develop the rate relation are described in the next subsection. Each method has certain advantages and disadvantages.

DEVELOPMENT OF THE RATE EQUATION

The importance of developing a kinetic rate equation on the overall process design cannot be overemphasized. The reaction rate not only will determine the size and number of reactors required in a continuous back-mix operation (and also batch operation) but also will affect the ability to design and control the operation to meet the proper pH guidelines. The potential impact of an oversized or undersized reactor on the operation will be discussed in Section 4.

Differential Analysis

The relation between residual reagent, B, and time is shown as a solid line in Figure 14. The following numerical approximation is sometimes used as an alternative that eliminates the graphing steps:

$$dB/dt \approx \Delta B/\Delta t \approx (B_1 - B_2)/(t_1 - t_2) \tag{3}$$

If the intervals $(B_1 - B_2)$ are small enough, adequate accuracy is obtained, but a more accurate method would be to draw the graph of B versus t and measure the slopes with a rule. The ΔB and Δt approximations are often used because values of B may vary over a wide range of concentrations (e.g., pH values of 1 to 7). The value of the slope at several values of B is determined, and the data are fitted to an equation relating reaction rate to residual concentration of reagent.

Neutralization reaction rates for specific acid-alkali systems can often be expressed as follows:

$$r = dB/dt = kB^n \tag{4}$$

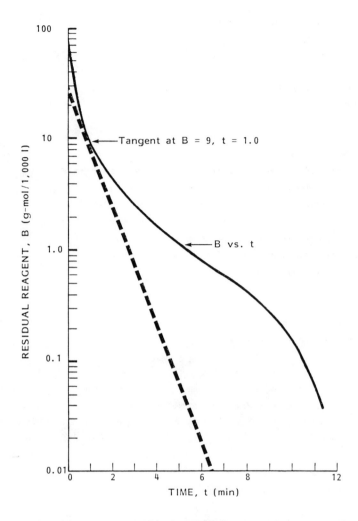

Figure 14. Semilog plot of conversion of reagent B as a function of time in a batch reactor.

where r = reaction rate at some specific time during the reaction
k = rate constant
n = constant (order of the reaction; often between 1 and 2)

Taking the log of both sides of the equation results in the following expression:

$$\log (dB/dt) = \log k + n \log B \qquad (5)$$

where B = residual reagent (mol/l)
dB/dt = rate of reaction (mol/l-min)
k,n = constants

Thus, if the reaction rate equation conforms to Equation 5, a plot of r versus B will yield a straight line on log-log graph paper, and values of k and n can be obtained from the graph.

This method, called the differential analysis method, has some disadvantages. First, a general kinetic model was assumed to fit the data when the following relation was assumed:

$$r = dB/dt = kB^n \ (-kB^n \text{ if B is disappearing}) \qquad (4)$$

Forcing this fit with the "best" straight line through the points, however, may not be a good fit and may introduce an error in design.

Integral Analysis

The integral method of analysis, based on the same rate relation as above, is similar in procedure except Equation 4 is integrated first for the general case. It is more difficult in principle but more accurate because the original experimental data are used directly before slopes are taken with a straight edge.

Integrating Equation 4 yields the following:

$$\int_{B_1}^{B} dB/B^n = \int_{0}^{t} k\,dt$$

$$B^{1-n} = (1 - n)(kt) + B_1^{1-n} \qquad (6)$$

where B_1 = initial reagent addition

The following iterative procedure is used to solve Equation 6 for n and k. A trial value of n is selected and Equation 6 is solved for k for each of several values of B and t. When k is relatively constant for each of the values of B and t [i.e., when a plot of $(B^{1-n} - B_1^{1-n})$ versus t is a straight line], the correct value of n has been found.

The answers in the differential and integral methods should be relatively close depending on how accurately the slopes were measured with the rule. The integral method can require a complicated trial-and-error solution.

Statistical Analysis

In the statistical approach to obtaining a reaction rate equation, the original concentration and time data are fitted to an assumed form of rate equation by statistical regression analysis techniques.

It is customary to limit the field to three types of possible kinetic rate equations--power function, exponential, and hyperbolic--which are shown by Equations 7, 8, and 9, respectively:

$$y = ax^b \qquad (7)$$
$$y = ab^x \qquad (8)$$
$$y = (a + b)/x \qquad (9)$$

Statistical analysis is mathematically a better method than either differential or integral analysis because it statistically fits the experimental data to the best equation. However, because commercial grade reagents contain impurities that sometimes amount to several percent, the less active fraction should be considered in the statistical analysis.

The statistical approach to the laboratory data can be simplified by access to a time-sharing computer that can fit data to the best curve in minutes. If hand calculations would take too long, a differential or integral analysis will be adequate. Using statistical regression analysis fits the data of residual base, B, versus time, t, to each of the three equations to obtain B as a function of t:

$$B = f(t) \qquad (10)$$

The equation with the best correlation factor is then differentiated to give dB/dt:

$$dB/dt = f'(t) \qquad (11)$$

If Equation 11 is explicitly in t, substitute B from Equation 10 for t in Equation 11. Although not necessary, using the rate equation, dB/dt, in reactor design facilitates the calculations. Now the rate equation is in the form

$$dB/dt = g(B) \qquad (12)$$

The three methods of establishing the rate equation for neutralization of a wastewater sample are illustrated in Example 3 of the appendix.

CONTROL OF PRECIPITATE PROPERTIES

Batch Neutralization

The control of precipitate properties to achieve dense dewaterable sludges can be visualized as the adjustment of reaction conditions to increase the particle size of the precipitate. In actuality, particle size distribution may also play a secondary role. Particle size can be increased by using seed crystals in conjunction with adjusting process conditions to promote the deposition of reaction product precipitate on the seed nuclei. Equation 13 quantifies for calcium sulfate precipitation the mechanism for verification of process performance through the design of reaction conditions into conformance with deposition kinetics. The adjustment of reaction conditions into conformance with deposition kinetics involves the following redefinition of the terms of Equation 13.

$$R = kM(r - 1) \tag{13}$$

where R = rate of addition of calcium ion or sulfate ion bearing reagent (g-mol/min)

k = rate constant (g-mol/g-min)

M = arbitrarily selected $CaSO_4 \times 2H_2O$ seed crystal dosage (g)

r = relative saturation--a value of 1.25 is suggested for initial experimental design

The use of Equation 13 should simplify the extent of experimental study required to design neutralization processes with controlled precipitate properties. The process objective is achieved by adjusting the reagent feed to agree with calcium sulfate deposition kinetics.

Scale formation is caused by crystal growth on the surfaces of pipes and equipment. Scale deposits from wastewater neutralization often consist of sulfates, sulfites, and carbonates of calcium with possible contamination by reaction products containing iron or aluminum. Fluorides and phosphates of calcium also have the potential to form scales. Because calcium is the common ingredient, monitoring the calcium concentration of effluents from developmental experiments as a function of time will indicate the potential for scale formation. The relief of postneutralization supersaturation can be accomplished by increased seed crystal dosage or by extension of the reaction period without addition of reagent to extend the contact of the effluent with the seed crystals and precipitate product [13]. The suggested developmental experimentation will enable the evaluation of the process conditions required for control of scale formation in batch neutralization.

Continuous Flow Neutralization

The design of developmental experiments for the evaluation of continuous flow processing requires the consideration of reactor purging during processing. The following relations permit the estimation of calcium sulfate deposition on seed crystals for steady-state neutralization of sulfuric acid solutions in completely mixed reactors:

$$F_i C_i = (F_i + F_m + F_b)C_e + V(dC/dt) \tag{14}$$

$$M = F_m C_m V/(F_i + F_m + F_b) \tag{15}$$

$$R = kM(r - 1) \tag{16}$$

where F_i = acid inflow (l/min)

F_m = seed crystal slurry inflow (l/min)

F_b = calcium-bearing base inflow, often negligible (l/min)

C_i = inflow acid concentration (g-mol/l)

C_e = effluent acid concentration (g-mol/l)

C_m = $CaSO_4$ X $2H_2O$ seed crystal slurry concentration (g/l)

R = rate of deposition on seed crystals (g-mol/min)

k = rate coefficient (g-mol/g-min)

M = $CaSO_4$ X $2H_2O$ seed crystal mass (g)

r = relative saturation, dimensionless, with value of 1.25 suggested for initial experimentation

V = reactor volume (l)

dC/dt = rate of neutralization (g-mol/l-min)

An expression is obtained for the required reactor volume by simultaneous solution of Equations 14 through 16:

$$V = \frac{F_i C_i - (F_i + F_m + F_b) C_e}{\frac{kC_m F_m (r-1)}{F_i + F_m + F_b}} \qquad (17)$$

Equation 17 should help translate the results of batch experimentation to the continuous flow mode. The batch experiments would evaluate F_m, C_m, r, and, if necessary, F_b. Improved performance in terms of sludge properties from continuous flow neutralization can be obtained by using multiple reactor stages to achieve increased particle size in the sludge wasted from the final stage and to provide better agreement with the batch experiments [14].

Increased control of scale formation (desupersaturation) can be achieved by bringing seed crystals into contact with the effluent in an additional reactor stage or in a reactionlike clarifier without the addition of reagent. The expected performance of the additional stage can be estimated from the results of batch studies translated to continuous flow by the Jones method described in Section 4.

4

Process Design for Neutralization and Precipitation

The heart of a wastewater neutralization and precipitation system is the pH control process, which must bring the wastewater pH to the level required to precipitate the optimal quantity of contaminant metal salt. This section presents fundamentals relating to the design of the components of the pH control process.

REACTOR SIZING AND GEOMETRY

The design equation that follows for an in-line reactor [15] is developed from a mass balance under steady-state conditions:

$$V/F = \int_{B_1}^{B_2} (dB/r) \tag{18}$$

where V = volume of reactor
- F = mass feed rate of reacting species in the wastewater
- B = conversion of reacting species in terms of mass of species reacted per unit mass of species initially present (subscripts 1 and 2 refer to the conditions at the entrance and exit of the reactor, respectively)
- r = rate of disappearance of reactant per unit volume

The volume of a stirred tank reactor operating at steady state is calculated from Equation 19 [15]:

$$V/F = (B_2 - B_1)/r \tag{19}$$

Because the reaction rate is constant throughout the reactor, no integration is required.

For a batch operation, it is necessary to establish a cycle time, θ, that consists of reaction time, time for filling and emptying, and any other operation, such as sedimentation, that may be carried out in the reactor. The reaction time, t, is calculated from Equation 20 [16]:

$$t = m/m_t \int_{B_1}^{B_2} \rho(dB/r) \qquad (20)$$

where m = the mass of reactant species
m_t = total mass of reaction mixture
ρ = density of the reaction mixture

This equation is for isothermal operation. Although neutralization reactions are normally exothermic, in wastewater systems the large volume of water--compared with the quantity of reacting species--tends to minimize heat effects. For details on the design of batch reactors operated under nonisothermal conditions, refer to J. M. Smith's <u>Chemical Engineering Kinetics</u> [15].

The minimal reactor volume, V, is determined by multiplying the average wastewater flow rate, F_{av}, by the cycle time, θ:

$$V = F_{av} \times \theta \qquad (21)$$

It is wise to have excess capacity in the reactor to allow for contingencies such as abnormally high waste volumes and equipment malfunctions. Occasionally, an additional reactor and a holdup tank are employed.

In a semibatch operation, the mass flow rate into the reactor is different from the mass flow rate out of the reactor. Equation 22 shows the form of the design equation for this situation [15]:

$$F_o(W_r) - rV = M_r [d(VR)/dt] \qquad (22)$$

where F_o = the total mass feed rate of reagent solution
W_r = the mass fraction of reagent in the solution
M_r = molecular weight of reagent
R = molal concentration of reagent in the reactor

This equation is solved by numerical integration. The reader is referred to Chemical Engineering Kinetics [15] for details.

Geometry is a factor in the design of tank reactors, but it is of little significance with in-line reactors where the reaction vessel is a pipe or facsimile thereof.

Tank neutralization reactors normally are cylindrical or square with a vertical axis. In standard designs, the liquid depth is equal to the diameter or width of the vessel. The impeller is usually mounted overhead with the drive shaft coincidental with the axis of the vessel (see Figure 12 in Section 3). Cylindrical vessels are baffled to control swirling; baffles are unnecessary in square tanks.

The standard geometry for agitated vessels was illustrated in Figure 12. The ratios of dimensions relative to standard geometry are shown in Table 4.

Designers of agitated reaction vessels usually pattern the reactors after the standard geometry to draw upon published correlations of agitator performance. Departures from standard geometry of ± 20 percent are often acceptable, although a change from $Z:D = 1:1$ may require compensation for power consumption. Design procedures have been developed for more substantial modifications of the standard geometry, such as deep tank vessels that employ multiple agitators. The reader is referred to literature sources for more comprehensive information regarding particular applications [16-21]. Technical information is also available from vendors.

TABLE 4. RATIOS OF DIMENSIONS TO STANDARD GEOMETRY FOR AGITATED VESSELS

Ratio	Propeller	Flat turbine	Pitched turbine
Liquid height to reactor diameter (H:T)	1:1	1:1	1:1
Agitator diameter to reactor diameter (D:T)	1:3	1:3	1:3
Baffle width to reactor diameter (B:T)	1:12	1:12	1:12
Distance between bottom of reactor and centerline of agitator to agitator diameter (Z:D)	1:1	1:1	1:1
Agitator blade width to agitator diameter (W:D)	NA[a]	1:5	1:8

[a] NA = not applicable.

MIXING SYSTEM DESIGN

Energy Requirements

The power dissipated during mixing is a function of the turbine diameter, its shape and speed, and the specific gravity of the fluid:

$$P = (N_p n^3 D^5 \rho)/g \qquad (23)$$

where P = power delivered to the fluid (m-kg/s)

N_p = power number, dimensionless

n = impeller rotational speed (r/s)

ρ = fluid density (kg/m^3)

D = impeller diameter (m)

g = gravitational constant (9.8 m/s^2)

The power number is a function of the Reynolds number (Figure 15). The Reynolds number, N_{Re}, for a fluid being mixed is given by

$$N_{Re} = (\rho n D^2)/\mu \tag{24}$$

where ρ = fluid density (kg/m^3)
  n = impeller rotational speed (r/s)
  D = impeller diameter (m)
  μ = fluid viscosity (kg/m-s)

The power required to achieve a given suspension often cannot be predicted with any reasonable accuracy and must be factored from small studies of the specific agitation problem. Some approximate correlations of agitation requirements for suspension of defined solids in vessels of different size have been presented by Gates et al. [16].

Dead Time

An important function of a mixing system in neutralization and precipitation processes is the minimization of the time between the addition of reagent and the first observable change in effluent pH or other process response. This interval is called dead time. For good process control, dead time should be less than 5 percent of the reactor residence time, V/Q, where V = volume of the reactor, and Q = volumetric flow rate of the wastewater through the reactor [22].

D.L. Hoyle [23] developed an empirical correlation of reactor dead time, vessel capacity, and agitator pumping capacity (Figure 16). The agitator pumping capacity is related to the agitator size and speed, as shown by Equation 25:

$$Q_a = N_Q n D^3 \tag{25}$$

where Q_a = pumping rate (m^3/s)
  N_Q = impeller discharge coefficient

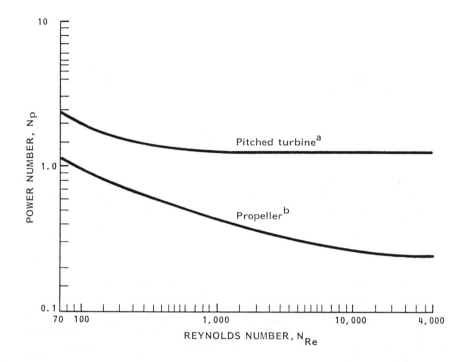

[a] R. L. Bates, P.L. Fondy, and R. R. Corpstein, Industrial Engineering Chemical Process Design Development, 2(4):311, 1963.

[b] J. H. Rushton, E. W. Costich, and H. D. Everett, Chemical Engineering Progress, 46:467, 1950.

Figure 15. Relation between power number and Reynolds number for pitched turbine and square pitched propeller agitation in vessels of standard geometry.

n = agitator speed (r/s)

D = impeller diameter (m)

Tank turnovers per minute = (Q X 60 s/min)/tank volume   (26)

For propellers with a pitch of 1.0 in baffled vessels, a value of 0.5 for N_Q has been suggested for design [19]. For pitched turbines with a width-to-diameter ratio of 1:8, N_Q is approximately 0.75 when the mixing Reynolds number is greater than 4,000 [21].

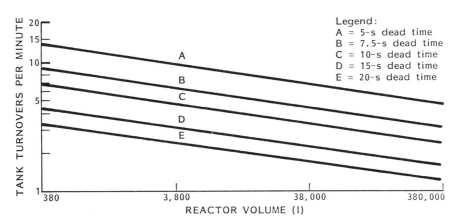

SOURCE: D.L. Hoyle, "Designing for pH Control," Chemical Engineering, 83(24):121-126, 1976.

Figure 16. Relation of mixed reactor dead time, reactor volume, and impeller flow.

Scale-Up

The scale-up of agitated vessels has been reviewed by Rautzen et al. [24]. The following relation was proposed for the scale-up of geometrically similar vessels agitated with turbine agitators:

$$n_2/n_1 = (D_1/D_2)^m \tag{27}$$

where n_1, n_2 = impeller rotational speed of model and prototype, respectively
D_1, D_2 = impeller diameter of model and prototype, respectively
m = scale-up exponent (m = 2/3 suggested for mass transfer; m = 3/4 suggested for solids suspension; m = 1.0 suggested for liquid motion)

For scale-up of equal blend time an m value of 0 is indicated. Such a scale-up is impractical because of large equipment size and high power requirements. The volumetric ratio for scale-up is preferably less than 2,000 and often ranges between 100 and 1,000.

Scale-up of neutralization reactors that relates process response to agitation intensity and vessel size has been developed [18]. For insoluble reagents, such as lime or limestone, the reaction rate frequently stabilizes when agitation is sufficient to keep all solids in suspension. The agitation required to maintain suspension of defined particles in reactors of various sizes is discussed in an article by Gates et al. [16].

The scale-up of neutralization and precipitation process reactors is frequently complex because:

- Liquid motion is required for reaction.
- Solids suspension may be required for accommodation of grit or particulate precipitates.
- Mass transfer may be required for dissolution of reagent.
- Constant blend time may be indicated for contact nucleation.

In addition, flocculent precipitates may be present that can undergo irreversible dispersion if subjected to excessive turbulence. This situation requires compromise in the design and scale-up.

An approach to the problem of complex scale-up involves the performance of process development studies in pilot-scale model reactors of different size, with one model possessing from 10 to 50 times the volume of the other. The results of the process development studies from the model reactors can be analyzed for the effects of scale-up on process variables. After scale effects have been identified, it may be possible to project the effects to the prototype reactor in conjunction with a controlling scale-up exponent.

Khang and Levenspiel [25] developed a dimensionless correlation for estimation of mixed reactor blend time:

$$K = [n(D/T)]^{2.3}/N_{MR} \qquad (28)$$

where K = nonhomogeneity amplitude decay rate constant (s^{-1})
  n = impeller rotational speed (r/s)
  D = impeller diameter (m)
  T = vessel diameter (m)
  N_{MR} = mixing rate number

The mixing rate number is a function of the mixing Reynolds number and is obtained from Figure 17 for propellers and pitched turbines operating in baffled vessels of standard geometry.

The blend time, t_b, is obtained from

$$t_b = (1/K)[\ln(2/A)] \qquad (29)$$

where A = amplitude of residual concentration fluctuations or degree of blending

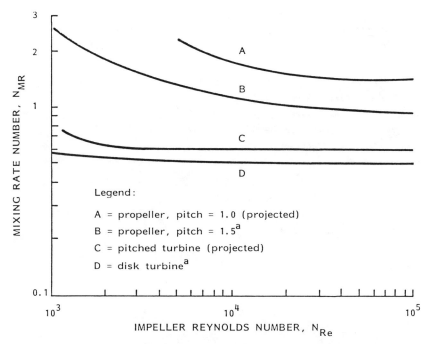

[a] S. J. Khang and O. Levenspiel, "The Mixing-Rate Number for Agitator-Stirred Tanks," Chemical Engineering, 83(21):141-143, 1976.

Figure 17. Relation between mixing rate number and Reynolds number for turbine- and propeller-agitated vessels of standard geometry.

It is suggested that the retention time, θ, of the scaled-up vessel be increased by the difference in blend times between the prototype vessel and the scaled-up vessel for a blending effectiveness of 98 to 99 percent (A = 0.01 to 0.02).

Failure to provide volumetric compensation for differences in blend time can lead to a scale-up that employs overagitation to compensate for nonhomogeneity. Overagitation wastes power and may cause permanent disintegration of floccules. References 21 and 24 provide more comprehensive information on scale-up of mixed reactors.

Flocculation

A key objective in flocculation is the maintenance of controlled agitation conditions so that particle contact is promoted without the presence of shear levels that can disrupt the delicate floc. Efficient flocculation has the potential for significant improvement in the performance of coagulation processes and metals precipitation processes [26].

Agitation levels for flocculation are usually expressed in terms of the agitation velocity gradient, G, which is defined as:

$$G = (\text{power/viscosity} \times \text{volume})^{1/2} = [(Nm/s)/(Ns/m^2)m^3]^{1/2} \tag{30}$$

where N = force (N)
m = distance (m)
s = time (s)

Values of G from 20 to 90 s^{-1} are typical for flocculation units. Tapered flocculation employs high entrance values of G and lower values as the flow progresses to the exit. Values of G X t (where t = seconds of residence time) ranging from 30,000 to 150,000 are commonly employed for flocculation in domestic water treatment. Flocculator retention times of 5 to 30 min are typical. Experimentally derived values of G and G X t are advocated for industrial waste applications.

CONTROL SYSTEM DESIGN

Control system design for batch neutralization processes can be quite simple because the system is conceived to perform a large-scale titration automatically. Batch systems can employ on-off control via solenoid or air-actuated valves. If the titration curve slope at the desired pH (set point of controller) is steep, the control system must be carefully engineered to prevent overshoot. Slow addition of reagent helps but can lead to impractically long treatment periods. An alternative is to use rapid reagent feed at the beginning of the titration and have an alarm switch automatically change to a small valve or feed pump to complete the titration to the set point. Metering pumps are preferable for reagent feeds of less than 1.89 l/s (0.50 gal/s). A cycle timer may be inserted in the reagent feed circuit to allow the pH to stabilize between reagent additions. Control valves with proportional plus derivative control can also prevent overshoot in batch neutralization performance. Supplementary controls for batch processes can include level controls, timers, and the like to automate the various operations in the process cycle [22].

Control system designs for continuous flow neutralization systems must cope with the problem of maintaining output pH within specification during operation on feeds that have fluctuating flows and concentrations.

Range of Operation

Mechanical limitations restrict the range of flow rate within which a control valve may operate. The range of operation is usually about 50 to 1. This means that flow rates of less than 0.02 times the maximal flow rate through the valve cannot be obtained with any degree of certainty. A single reagent flow control valve is likely to be inadequate, considering that pH control often requires a thousandfold or greater range in reagent delivery rate capability. Sequencing a small control valve to work in conjunction with a larger control valve may have a practical range of operation of about 1,000 [22]. Each valve operates independently to cover a selected range of pH adjustment. The system's theoretical range of operation is the product of the individual valves' operational ranges.

Control valves differ in range of operation by type. For slurried reagents, ball valves or diaphragm valves are usually employed, with the latter having the lower range of operation. Proportioning feed pumps and dry feeders typically have a range of operation of 20 to 1, although a higher range of up to 100 to 1 is available from equipment with variable stroke and variable speed.

Attenuation

The pH control requires a finite amount of time to respond to changes in pH caused by system, mixing, and reaction lags. Sudden changes in pH cannot always be responded to in time to prevent some of the wastewater that is outside acceptable limits from passing through the system. It is desirable, therefore, to provide attenuation within the system to slow down the rate of change in pH. Attenuation can be provided by wastewater equalization facilities or surge vessels.

Attenuation also can be provided within the neutralization system itself. Continuous back-mixed reactors with a ratio of dead time to retention time between 0.05 and 0.10, preferably near the lower value, provide attenuation. Good mixing ensures that the pH sensor will detect changes in wastewater pH readily, which helps the controller respond rapidly to changes in pH.

Aspects of Reactor Design

Certain aspects of reactor design are required for good process control for batch processing. These consist of sufficient capacity to contain and process the flow and sufficient agitation to blend the reagent with the waste. These will enable the sensor to monitor a uniform mixture. With continuous flow processing, the reactor design aspects of process control are far more significant. The reactor has a retention time that can be calculated as volume divided by flow. The reactor has a dead time that is a function of tank volume and agitation efficiency, as indicated in Figure 16.

Good agitation depends on intermixing to mix the reagent with the wastewater feed and back mixing to maintain uniformity of reactor contents. Inlets and outlets of reactors should be arranged to pass the flow through the agitator (Figure 18). A change in the direction of the inlet channel, such as a 90° bend, will assist with the intermixing of reagent with the flow. Flow-proportional reagent addition is desirable for feeds with significant flow fluctuation and finite reaction time. Increased reactor size is a mixed blessing because it increases both conversion in sluggish reactions and equalization of short period input concentration fluctuations, but it also adds dead time to the control loop.

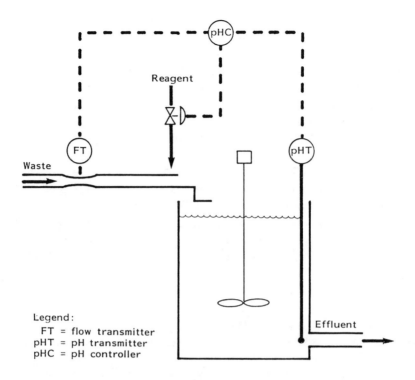

Figure 18. General purpose feedback controlled mixed reactor.

A smoothing vessel is frequently employed as an aid to neutralization and precipitation processes. The vessel may be agitated or unagitated, but it is not provided with a reagent feed or a control system. It has the function of attenuating the fluctuations in pH that pass the neutralization system or that are generated by the limit cycling of the control system. In many cases the smoothing vessel is a settling tank that has the primary purpose of separating suspended solids, including precipitated reaction products and unreacted reagent. With slow-dissolving reagents, the separation stops the reaction.

The effective retention time, $m\theta$, of smoothing vessels may approach 0.9θ with mild agitation, 0.8θ with mixing nozzles, and about 0.7θ with no induced mixing but designed to limit short circuiting. With settling tanks the effective retention generally exceeds 0.60θ, but the effects of separation of residual reactant on pH drift require experimental evaluation.

Because smoothing vessels are outside the control loop of the neutralization process, dead time associated with smoothing vessels has no effect on the control system.

The following relation has been proposed for estimating the performance of smoothing vessels [27]:

$$\frac{dN_e}{dN_i} = \sqrt{\frac{1}{1 + (2\pi m\theta/T)^2}} \tag{31}$$

where m = coefficient of effectiveness of retention time, dimensionless

θ = retention time, V/F (s)

T = period of sine wave concentration surge (s)

N_i, N_e = normality of residual reagent in the inlet and outlet streams, respectively

For strong-acid, strong-base neutralization processes with total dead time equal to 0.05 times the retention time, Hoyle [23] has suggested a one-reactor system for feeds with pH fluctuation of 4 to 10, a reactor plus a smoothing tank for feeds with pH fluctuations of 2 to 12, and two reactors plus a smoothing tank for feeds with a pH of less than 2 or greater than 12.

In the final analysis, however, the nature of the wastewater feed fluctuations warrants consideration. Mixed reactors under feedback control exhibit a natural period at which control performance is poorest. Feeds containing waveform fluctuations in flow or concentration corresponding approximately to the natural period receive inferior processing. For mixed reactors under conventional three-mode control, the natural period can be estimated as 4 times the total dead time [27]. Improved control performance is attainable by applying equalization to attenuate feed fluctuations with periods shorter than 4 times the natural period of the reactor system.

Fluctuations in wastewater concentration are usually of much greater magnitude than fluctuations in flow. Consequently, neutralization reactors are sometimes designed to control only concentration fluctuations, with the assumption that the lesser flow fluctuations will be accommodated as a matter of course. The assumption may be erroneous if flow fluctuations are substantial because a change in reactor retention time can offset process control if reaction rates are finite. A flow sensor in the control system will improve response to flow fluctuations, but the best prospects for reaction control are provided by flow equalization in conjunction with flow sensing. The equalization is designed to attenuate brief fluctuations that cause excessive dead time, and the flow sensor input will enable the control system to accommodate the longer fluctuations that require excessive equalization capacity.

Many alternatives are available for feedback control of neutralization reactors. The basic concept is a single-stage, back-mixed reactor with intermixing of reagent with wastewater inflow and control by pH sensor (Figure 19a). The system is applicable to essentially uniform flow situations and limited acid/base concentrations in the wastewater. Modifications to upgrade the concept might consist of the provision of split feed (dual feed of acid or base reagent) to enable the accommodation of either acidic or basic wastewater discharges. The inclusion of equalization plus flow sensing would improve the capability of the basic system to process fluctuations in wastewater flow and concentration. The provision of a smoothing vessel after neutralization will assist in the stabilization of the effluent and will extend the overall range of operation of the system.

Multistage reactors (e.g., Figure 19b) are required to process concentrated wastes. Staging effects an upgrading in operational range equivalent to the product of the components' ranges of operation. A number of stages, for example, six, may be required for the neutralization of concentrated wastes with reagents possessing sluggish reaction rates. Cascade control, in

Process Design for Neutralization and Precipitation   71

(a)

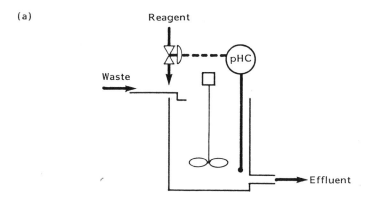

(b)

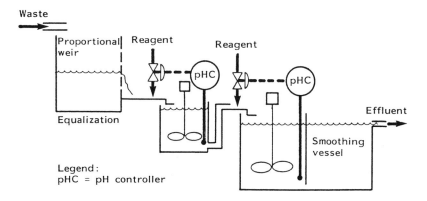

Figure 19. Alternative concepts for feedback control: (a) single-stage back-mixed reactor and (b) multiple-stage reactor.

which the set point of the first reactor is subject to adjustment by the controller governing the second reactor, is a modification that offers potential for improved performance in difficult situations [23].

REACTOR STAGING

Staging refers to the use of two or more continuous back-mixed reactors in series so that the reaction does not go to completion until the last stage. Staging may be used to limit the size of individual reactors when long residence times are required. Process control is another factor that may favor the application of staged reactors. As mentioned earlier, a ratio of dead time to retention time of close to 0.05 was recommended for effective reaction control by instrumentation [27]. Because large reactors inherently incorporate substantial dead time for blending, better control is obtainable by providing the necessary capacity in several staged reactors rather than in a single reactor of comparable overall size. In addition, staged reactors provide the opportunity for multiple instrumentation, which offers great potential for upgraded process control.

The design of staged reactors usually commences with an assumed retention time based on the performance observed in the experimentation. The retention time per stage may be as short as 3 min for extremely rapid reactions or up to 15 min for reactions of moderate velocity. Longer retention times may contradict the objective ratio of dead time to retention time.

To determine the approximate number of stages, use the reaction rate equation to calculate values of the residual reagent concentration in successive reactors with a retention time, θ, equal to approximately 10 min until the value of the residual reagent concentration corresponds to a pH value within the tolerable limits. (In each calculation, the residual reagent concentration of one reactor becomes the inlet reagent concentration of the next reactor.) Mathematically, we have:

First stage: Disappearing reagent B ($r = -dB/dt$)

$$B_2 = r\theta + B_1 \tag{32}$$

where B_1 = input reagent concentration (mol/l)
B_2 = output reagent concentration (mol/l)
r = dB/dt (mol/l-min)
θ = reactor detention time, V/F (min)

For the second stage, B_2 is now B_1 in Equation 32. For the nth stage, repeat the calculations until the calculated value of B_2 is within the corresponding pH range.

Example 4 in the appendix illustrates the use of this procedure to design staged reactors for the wastewater in Example 3. The reaction rate equation developed by the statistical technique will be used.

Graphical Back-Mix Design Procedure

R. W. Jones [28] developed a method for designing staged reactors that uses the graph of reaction rate versus residual reagent directly and does not require a determination of the reaction rate equation. Rearranging Equation 32, the following graphical procedures become evident:

$$(B_1 - B_2)/\theta = -r \tag{33}$$

$$(1/\theta)(B_1 - B_2) = -r = dB/dt \tag{34}$$

Then
$$y = -r = dB/dt \tag{35}$$

$$y = (1/\theta)B_1 - (1/\theta)B_2 \tag{36}$$

If Equations 35 and 36 are graphed, the point of intersection is a solution to Equation 32. This is how the Jones method works. Note that Equation 36 is a straight line on linear paper with a slope of $-1/\theta$ and a y-intercept of $(1/\theta)B_1$.

Measure the slopes at various values of B on the graph of B versus t; then plot the slopes dB/dt (-dB/dt if B is disappearing) versus B. Graphically, the process looks like Figure 20. Certain assumptions are made with this graphical procedure. First, the composition in each stage (vessel) is uniform throughout. Second, the rate of change (reaction rate) of B with respect to time is the same as in a batch operation. Third, the relation of B versus t is continuous over its domain, and no change of slope (zero order) occurs over the range of variables.

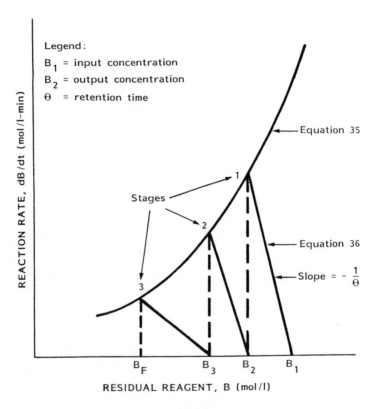

Figure 20. Graphical projection of batch reactor results to continuous flow reactor performance by the Jones method.

Now a line can be drawn through B_1, as described in Equation 36, that has a slope of $-1/\theta$, and by specifying a θ (such as 10 min) it can be seen where that line intersects the curve (Equation 35). This is the first stage. Dropping a vertical line to the B axis, where $-dB/dt = 0$, yields the value B_2, which is the concentration of reagent in the effluent from the first reactor, with a 10-min retention time. Continuing until the desired B value is reached (the first value which falls within the desired pH) will indicate the number of stages with a retention time of 10 min. It should be noted that changing the retention time, θ, for each stage is possible when a straight line through the B value is drawn with a different slope $1/\theta$.

Example 5 in the appendix illustrates the use of the Jones method for designing staged reactors for the wastewater of Example 3. Note that the described technique is performed on rectilinear paper. Because neutralization involves large concentration ranges, a log-log plot is usually more convenient. The procedures are the same except Equation 36 no longer can be drawn as a straight line on log-log paper and, therefore, must be plotted at various B values for each y chosen. The point of intersection with the rate equation (Equation 35) will give the same answer. Whether the slope of Equation 36 is positive or negative depends on whether the rate describes the appearance or disappearance of the reagent. It will become clearer if Equations 35 and 36, when graphed, diverge instead of intersect. In this case, there is a sign error. Example 5 illustrates the technique more clearly.

A tabulation of the results of analysis by the differential, statistical, and graphical methods is given in Table 5. The results indicate adequate agreement between the methods; therefore, the selection is largely a matter of preference. As a rule, the Jones method is most convenient for routine analysis because it does not involve a search for a kinetic relation. The assumption of retention time facilitates the analysis, inasmuch as solving for retention time involves an iterative procedure. The differential and statistical methods are more amenable to incorporation in analysis of process dynamics, including the evaluation of dead time associated with reaction rate lag.

TABLE 5. RESULTS OF ANALYSIS BY THREE METHODS

Stage	Differential		Statistical		Jones	
	B_2	pH	B_2	pH	B_2	pH
1	7.12	1.45	7.83	2.38	7.6	2.40
2	1.23	4.47	1.62	4.0	1.25	4.65
3	0.34	8.11	0.30	8.25	0.2	8.65

All three methods of translating batch results to continuous processing incorporate such assumptions as complete mixing, instant dispersion of reactor feeds, and equivalent reaction rates. The assumptions hold adequately for most neutralization situations; however, the assumption of equivalent reactivity between batch and continuous flow systems may be of limited validity with reagents that are grossly impure or whose rate varies with pH, unless the analyst can synthesize a rate function that will compensate for the differences in chemical environment between the systems (e.g., reactivity may be a function of pH). Some limestone and waste product reagents fall into this category. The alternative is to employ continuous flow process development.

In the sample problems presented in Examples 1 through 5 of the appendix, each stage of the neutralization system was made equivalent in volume. Some authors recommend increasing the downstream size of reactors to obtain better process control. These design criteria could have been employed in the examples by decreasing the volume in the first stage and increasing it in subsequent stages.

Performance Projection

The reactor design described earlier is based on a particular set of wastewater conditions. During actual operation, wastewater characteristics will vary continuously, and it would be desirable to project the performance of the system under actual operating conditions to ensure that the effluent characteristics will remain within acceptable limits. Unfortunately, a general procedure for forecasting the performance of a neutralization system has not been developed.

F. G. Shinskey [27] proposed an analysis that provides guidance for projected performance for a number of strong-acid, strong-base situations. The predictive equations suggest that for a given acid/alkali system the sensitivity of the process to disturbances varies with the period of the upset, the ratio of dead time to retention time, and the departure between influent pH and the pH at the control point. Reduction in the difference between influent pH and set point, as achieved by staging reactors, is a dominant factor affecting performance. Adjustment of the buffer chemistry by change of reagents also offers strong potential for control of process sensitivity.

The performance of a second-stage reactor can be projected by using the value of the effluent pH from the first stage as the value of the influent pH for the second stage, with adjustment of the other terms of the equations into agreement with the second-stage reactor. Example 6 in the appendix illustrates some of the design concepts presented in this manual.

A substantial increase in complexity of analysis is involved in the prediction of the performance of neutralization systems when the assumption of "instant" reaction is an excessive departure from reality. Analog computers have been widely used for system simulation--particularly if nonlinear differential equations are involved. Digital computation using analog emulation is also applicable. Several emulation languages are available from time-sharing computer companies. A mathematical model for neutralization with a reaction time constant has been outlined by Moore [23]. The model is suitable for evaluation by analog emulation. Simulation of neutralization systems is recommended before installation.

The design of systems to process mixtures of strong and weak acids or bases that differ in composition from hour to hour poses particular problems relative to process control. A control system designed to dispense minute quantities of reagent to change a strong-acid, strong-base system from pH 5 to pH 7 cannot efficiently make the large additions of reagent necessary to neutralize a highly buffered weak-acid, strong-base system from pH 5 to pH 7. This type of situation requires problem-specific engineering of the neutralization system using such techniques as component equalization, reagent selection, supplementary staging, and adaptive control. Continuous flow pilot-plant testing using frequency response analysis is suggested for performance projection.

Back-Mix Reactor Design Theory

This discussion concerns how process variations affect the actual plant operation and how they affect and relate to all the experimentation and pilot-plant work that lead up to a final design. A better understanding of these relations will make the final design more responsive to those variables that affect the success or failure of the process.

Good pH control is difficult--more difficult than the control of most other process variables like flow, temperature, or level--largely because pH is a nonlinear function of concentration.

Let us assume that a waste effluent sample has been tested according to the procedures in this section and the resultant three-stage design is as illustrated in Figure 21. The dosage curves will simplify the discussion and understanding of how variables affect control and design.

The first stage has a retention time of 10 min and raises the pH from 1.5 to 2.9. The second stage brings the pH from 2.9 to 5 and also has a 10-min retention time. The third stage raises the pH from 5 to 8 and has a 10-min retention time in the reactor. Assume these pH values and retention times correspond to a rate equation and back-mix design based on that rate equation and that everything can be related back to the dosage curve that has been chosen for the design basis. The control can be described as feedback. Although each stage has the capacity to add reagent (ideally and according to the design equation), if the reactors could be designed according to the proper kinetic relationship, reagent addition would be required only in the first stage. If the exact amount of reagent could be added, it would bring

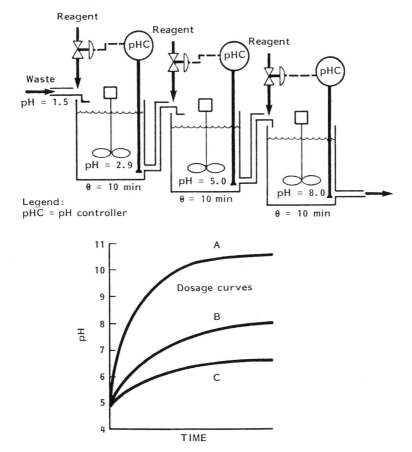

Figure 21. Three-stage reactor system for neutralization of acid waste.

the first-stage pH to the design value of 2.9; the other stages would merely permit the reaction to go to completion with no additional reagent required. Under ideal conditions, with all factors remaining constant, no control would be required. Additional reagent in the other stages is to compensate for <u>slight</u> variances in any stage. In practice, somewhat low dosages are added intentionally in the early stages to avoid overshoot.

`Considerations must be made for reactor efficiency, that is, how well the reactor size provides for complete mixing and the possibility of short circuiting. For reaction mixtures where the flow remains constant, little or no correction is required. The reactor capacity should compensate for any short circuiting if designed to have slightly more volume than is required to accomplish the required pH change. This compensation is accomplished in ordinary design because exact volumes are always increased to some standard size vessel. In neutralization this usually provides a comfortable margin of safety, if the initial kinetic model was accurate.

Once a dosage curve has been selected for operation, reactor retention times are fixed by reaction kinetics. The following paragraphs discuss the effects of large variations on wastewater flow rate and concentration.

Variable Feed

If the flow increased by a factor of 3, the pH controller would sense a pH below 5 at the effluent of the second stage. The reagent control valve would add reagent until a pH of 5 could again be attained in the exit stream. The increase in flow decreases the retention time in each stage. The residence time in each reactor is reduced to 3.33 min. The fact that the control system responded by adding more reagent to reach the pH control point of 5 in the effluent means that the system is no longer operating along the design kinetic curve. The system compensated for the <u>reduction</u> in <u>retention time</u> by adding a greater dosage of reagent to <u>decrease</u> the <u>reaction time</u> to reach a pH of 5. The system is following a different dosage curve and correspondingly different kinetics. Although the process stream in this stage is controlled to a pH of 5, more reagent has been added than is required to reach a final pH of 8 and the reaction will continue past the third stage. The final pH will be greater than 8 at some point downstream of the third stage. In actuality, what is required if flows should double or triple is an increase in reagent addition in proportion to the flow and compensation for less reaction time in this stage by providing additional time in the final stage. Also, the pH set point would have to be lowered to follow the design dosage curve. The addition of a flow sensor to the controller circuit and the design for maximal conditions is required to handle this situation.

In multistage neutralization, oversizing the first stage allows for more time than necessary and results in a shift to operation on a different dosage curve--in this case, to a lower curve. It is better design to allow for larger retention times in the final stage, which is the usual practice when flow variations are expected. Even if feedforward control is used to prevent overdosing the stream, a pH set point correction must be made. This will prevent the so-called dosage jump that systems tend to make when responding to large flow changes. If feedforward control is combined with a titration correction for pH, the system will provide better control if adequate volume is provided.

Variable Concentration

Concentration fluctuations are easier to handle than flow fluctuations. It is best to design for the highest expected concentrations so that the equipment will be slightly oversized and dosage jumps will be avoided.

Referring to the design in Figure 22, a lower acid concentration (higher pH) would signal the pH sensor that the reagent valve should be closed because less reagent would be required to change the pH to 5. No dosage jump is expected because reagent is being added proportionately to the dosage curve to achieve a final pH of 8. Thus, designing for high concentrations (low pH values) can result in larger reactors or additional stages.

Staging Control

Another advantage of staging is better control. Mathematically, a batch reactor can be simulated by an _infinite_ number of back-mix reactors. Batch reactors are easy to control, so it can be reasoned that a system mathematically equivalent to a batch system would also be easy to control. Although this reasoning is not a rigorous proof, it lends more credence to the staging philosophy when trying to control a waste effluent over a large pH span.

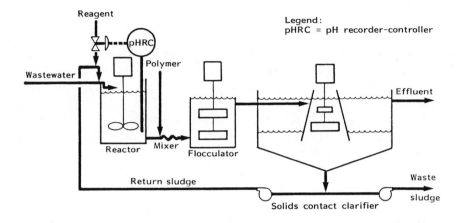

Figure 22. Process schematic for precipitation and separation of target substances.

The ramifications of oversized or undersized reactors should, in view of this discussion, be more apparent. Too much residence time in any stage except the last allows more time to achieve the desired pH and, depending on the control set point, could lead to an underneutralized effluent because of operation on a lower dosage curve. Conversely, excessively small reactors with insufficient residence time can result in excessive reagent addition and an overneutralized effluent.

It has also been suggested that in a staged reactor system the use of different sizes of reactors aids in control. Sometimes downstream reactors are made progressively larger to avoid amplification of disturbances at the natural periods of the first reactor. Figure 21 shows three-stage reactor systems with equal size reactors to evaluate kinetic modeling techniques on an equivalent basis. Using different volumes for reactor staging as a means of better control is related to the natural period of the system.

Control of Crystalline Properties

In the previous discussion it was assumed that the amount of reagent required to achieve the desired effluent pH was added to the reactor all at once. Often, however, the properties of the precipitate can be improved by controlling the rate of reagent addition and other process conditions.

Properties such as dewatering and crystal size can be controlled to the extent that, instead of the neutralization process creating a watery sludge that must be disposed of off site at considerable expense, sludge with excellent dewatering properties can sometimes be obtained and sent to recovery operations. High sludge disposal costs provide an incentive for applying methods of controlling precipitate properties.

Possible economic penalties associated with controlling the reaction at a slower rate are anticipated; longer retention time would be needed to complete the neutralization because pH objectives must still be met. Longer retention time requires larger reactors (or more stages) and more control, which contribute to greater capital expenditures. An economic evaluation must be made to determine whether the reduction in disposal cost offsets the additional costs and provides for an acceptable return on investment.

The experimental procedure for obtaining data for reactor design with a controlled reagent addition rate is similar to the procedure for adding reagent all at once. It is necessary, however, to be aware of how the laboratory procedures are simulated and to adapt these procedures to an analysis of the kinetics so that both crystal formation and pH are controlled.

Experimental Procedure

The first step in the experimental procedure is to derive reaction-rate/reagent-dosage curves by adding the amounts of reagent all at once. Once a dosage curve is selected, the reagent dose is added at various rates. Additional experiments may be performed that involve the addition of seed crystals or return sludge. It is obvious that the permutations of experimental conditions are numerous. In some cases, protracted maturation of the process is necessary for complete process development.

Each experiment involves the addition of the specified quantity of reagent at different rates. Along with dosage of recycle sludge, each rate addition is considered a run. After each run, the sample would be settled and an aliquot of the sludge would be tested for dewatering properties. This could entail many runs. Once the desired properties are obtained, the following parameters have been determined:

- quantity of reagent required to effect the pH change per liter of effluent
- rate at which this quantity must be added to the liter sample to effect both the desired sludge properties and final pH
- rate of sludge recycle and reaction conditions (e.g., temperature)

The use of this procedure is illustrated in Example 7 of the appendix.

Development studies for precipitation processes are relatively simple if the chemical characteristics of the wastewater are constant. Bench studies using techniques similar to those presented for neutralization are applicable. The expected process performance in terms of residual concentration of target substance must be verified by chemical analysis; it usually cannot be computed from theory. Analysis of settled and filtered residuals of target substance will assist in recognizing the presence of a colloidal suspension. If the residual concentration of metal salt in the supernatant liquid in the settled sample is significantly higher than in the filtrate of the filtered sample, coagulation with a polymer or metal salt may improve sedimentation performance materially. Precoat or another fine media filtration is an alternative means of separating unsettleable suspensions. After establishing the bench-scale batch performance, the process can be designed into a batch or continuous system by the procedures previously outlined for neutralization systems.

If the wastewater that contains target substances is a composite of independent sources, the treatability studies become more complex because of fluctuations in chemical characteristics. The process performance may be expected to differ with changes in target substances or with changes in accompanying chemical species. An assessment of expected process performance on fluctuating inputs can be obtained by statistical processing of bench-scale results in a manner analogous to the assessment of input flows and loadings presented in Section 2.

Techniques for controlling sludge properties discussed under neutralization are also applicable to target substance precipitation processes. In the case of crystalline precipitates, such as calcium fluoride and calcium phosphate, seed crystal nucleation processes are potentially effective. Control technologies applicable to amorphous sludges are applicable to metal precipitates. The effectiveness of these techniques should be explored in the process development studies.

The experimental procedures for controlling target substances are analogous to those described for neutralization, with the analytical focus being on the target substance(s) as well as on pH and sludge characteristics. That is, the experimental effort may evaluate alternative reagents, dosages, agitation intensities, reaction times, and the effects of process temperature, rate of reagent addition, and return sludge on process performance. The experimental procedure might include the following steps:

1. Examine the wastewater generating processes for elimination or reuse opportunities and conduct a survey to characterize the remaining discharge.

2. Collect representative samples of wastes from an operating process, if available, or prepare them synthetically, if necessary. Wastewater samples may require synthetic modification to reflect future production conditions. The variances of wastewater flows and characteristics are estimated from available plant records or by analogy to similar plants.

3. Analyze the samples for temperature, concentration of target substances, and reagent demand for control of pH (titration curve).

4. Prepare return sludge if it is to be used in the process development studies.

5. Place a sample in a bench-scale mixed reactor and add selected dosages of reagents and return sludge for pH adjustment and/or target substance precipitation.

6. Monitor the reaction variables as functions of time. Variables such as pH can be monitored by a sensor. Monitoring most variables will require the withdrawal of a sample for quick filtration to separate the soluble fraction and fixing by pH adjustment and/or dilution as required to stabilize the observed condition. The objective of the latter operation is to preclude any change in target variable during storage of the sample before analysis. The variables to be monitored include pH, target substance concentrations, and possibly residual reagent concentrations. For induced precipitation processes involving crystalline precipitates, the level of supersaturation is monitored.

7. Upon completion of the reaction period as evidenced by stabilization of pH or concentrations of key variables, transfer the reactor contents to a graduated cylinder or other settling vessel for separation of precipitate and observation of sludge volume. Samples of supernatant are periodically withdrawn for analysis of total and filterable target substance concentration and the sludge volume is recorded. The settling period is usually 18 to 24 h if sludge compaction information is to be obtained. The sludge suspended solids concentration is determined at the end of the settling period.

8. If data analysis reveals a substantial difference between settled and filtered concentrations, variables such as agitation intensity and reaction time may need to be evaluated, or a coagulation process may be required to render the precipitated material settleable. The coagulation process may consist of the addition of organic polymer or metal salt coagulant concurrently with reagent addition, or it may be an intermediate process between reaction and settling.

9. The design procedures for analysis of the experimental data, kinetic modeling, reactor sizing, and reactor staging are as outlined previously for neutralization systems.

Practical Application

Target substances can be precipitated by batch or continuous processing. Batch processing has the advantage of better control over continuous processing of effluent quality but requires larger areas. Continuous processing can often be accomplished in a process train similar to that illustrated in Figure 22. Laboratory or pilot-plant process development is recommended in conjunction with guidance from available literature [29]. As a rule, detention times range from 3 to 10 min in mixed reactors and 5 to 15 min in flocculation units.

Guidance for reagent selection is available from theory or from literature sources such as Table 3. Experimentation is necessary to verify process integrity. Polymer flocculants are often added in dosages ranging from 0.2 to 1 mg/l (0.3×10^{-4} oz/gal) to improve sedimentation performance. Polymer addition may be concurrent with reagent addition or between the reaction and flocculation operations. Flocculation detention time and agitation intensity are important process variables [26].

The results of a study by Thomas and Theis [30] on the effect of common foreign ions on the precipitation and sedimentation of various metals are of general interest. The study established that metal precipitation performance is adversely affected by the presence of phosphates, carbonates, and, to a lesser degree, sulfates. As illustrated in Figure 23, calcium hydroxide was more effective than sodium hydroxide as a precipitant for chromous ion in the presence of carbonates and sulfates. The improved performance of the calcium hydroxide was attributed to the precipitation of anions that form complexes with chromium and the contribution of doubly charged counterions to the solution. The results also suggest that hydroxy complexes of chromium were collected by calcium carbonate precipitate with calcium hydroxide treatment to high pH levels. The finding demonstrates that pure solution solubilities are of limited significance in industrial waste process development. The authors suggested segregating wastes containing phosphates and carbonates from metal-bearing wastewater flows to minimize the dispersion of metal salt precipitates.

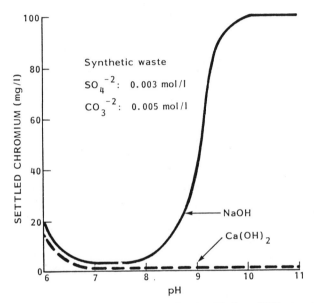

SOURCE: M. J. Thomas and T. L. Theis, "Effects of Selected Ions in the Removal of Chrome III Hydroxide," Journal Water Pollution Control Federation, 48(8): 2032-2045, 1976.

Figure 23. Comparison of performance of sodium and calcium hydroxides for precipitation and sedimentation of trivalent chromium from synthetic metal finishing waste.

There are numerous instances where process performance is better than theoretical performance. A massive precipitation of floc containing iron, aluminum, and calcium carbonate or phosphate usually will cause a generalized removal of metal salts. An EPA report [31] has proposed ferrous sulfate coagulation and lime overtreatment processes for the removal of complexed metals from electroplating rinse waters.

The precipitation of fluorides from electronics manufacturing wastewater was investigated by K. L. Rohrer [32]. The wastewater containing fluoride at 200 to 4,000 mg/l (0.026 to 0.512 oz/gal) was treated by calcium hydroxide in conjunction with calcium chloride. The results indicated that an optimal pH range of 6.5 to 9.0 existed for the precipitation of fluoride. The filtered effluent samples in the optimal pH range registered an average fluoride concentration of 17 mg/l (21.76 X 10^{-4} oz/gal) after 16 h of settling with polymer addition of 1.0 mg/l (1.28 X 10^{-4} oz/gal). Subsequent plant-scale test results agreed with the bench experiments except that the optimal pH range was 5.7 to 8.0. The use of calcium chloride in conjunction with calcium hydroxide promoted impressive fluoride removal in a process with a neutral effluent.

Several sources of information on the precipitation of target substances warrant recognition. Compendia on the performance of treatment technology for the removal of target substances have been published [29]. Reviews of technology applicable to electroplating wastes [33-35] and presentations of research results with a complete bibliography [31] are available.

CHEMICAL HANDLING AND STORAGE

Detailed information on chemical handling and storage is available from manufacturers and from the literature [36-39]. Design of a solid handling or slurrying system is best handled by an experienced vendor. References should be available from chemical suppliers.

Liquid Reagents

Some chemicals are shipped in concentrations that will freeze at temperatures slightly below room temperature. They are shipped warm in insulated transit tanks, must be kept warm during unloading operations, and must be maintained above their freezing point or diluted for storage. Crystallization and freezing temperatures vary with concentration. Concentrated chemicals often exhibit a perceptible vapor pressure of a solute (e.g., sulfur dioxide, hydrogen chloride) that may be toxic or corrosive. It follows that storage tanks for such chemicals require venting to the outside (with proper precautions) to prevent accumulations of vapor indoors. In addition, concentrated solutions frequently possess high viscosity and high density, which may require judicious allowances in the design of transport lines and pumps. Consideration should also be given to the chemical stability of reagents during long-term storage.

Bulk unloading facilities usually must be provided at the treatment plant. Rail cars, which sometimes are constructed for top unloading, require an air supply system and flexible connectors to displace the chemical pneumatically from the car. U.S. Department of Transportation (DOT) regulations concerning chemical tank car unloading should be observed. Tank truck unloading is usually accomplished by gravity or by a truck-mounted pump.

Industrial hazards may be encountered in handling liquid chemicals. A face shield and gloves should be worn around leaking equipment. Facilities for flushing the eyes and skin should be provided as part of systems for handling liquid chemicals in bulk. In certain instances, air masks or respirators must be available nearby.

Concentrated chemicals often possess high heats of hydration, so it is essential to follow the manufacturer's recommended procedures for dilution to avoid the possibility of an explosive reaction. Chemical storage tanks should be protected from pressure/vacuum situations during filling and emptying. They may require outside venting for release of toxic or corrosive vapors.

Storage tanks should be sized according to maximal feed rate, shipping time required, and quantity of shipment. The total storage capacity should be more than sufficient to guarantee a chemical supply while awaiting an order.

Gaseous Reagents

Carbon dioxide, chlorine, ammonia, and sulfur dioxide are available in cylinders for gas feeding. Considerable information on handling, storage, and feeding of gaseous reagents is available from suppliers and from vendors of feeding equipment.

Compressed gaseous reagents are shipped in bulk or in cylinders ranging in size from 45 to 900 kg (99 to 1,984 lb). Bulk shipments are made by truck or tank car in lots ranging from 900 to 1,650 kg (1,984 to 3,638 lb). Carbon dioxide, chlorine, ammonia, and sulfur dioxide should be considered toxic and corrosive. The safe handling and storage of pressurized gases requires specialized engineering of the systems and careful, knowledgeable operation. Predesigned package storage systems are available for some situations. Designs should be kept in strict conformance with industry codes and manufacturers' recommendations. Potentially hazardous situations should be analyzed thoroughly for each installation.

Polyelectrolytes

Polyelectrolytes are classified by their charge in the ionized form:

- Anionic polyelectrolytes--negative charges
- Cationic polyelectrolytes--positive charges
- Nonionic polymers (nonionic polyelectrolytes)--neutral charge

Polymers can be purchased in dry or liquid form. They are easily handled at the plant site and are nonhazardous. The usual protection from dust is required, and the storage facilities for the dry powders must be completely dry. Construction materials for storage and handling equipment are normally type 316 stainless steel or plastics, depending on the polyelectrolyte chosen.

Different polymers vary widely in characteristics. Manufacturers should be consulted for properties, availability, and cost of the polymer being considered. Bulk shipments usually are not desirable. Polymers are available in a variety of containers or package sizes.

General practices for storage of bagged, dry chemicals should be observed when storing polymers. The bags should be stored in a dry, cool area and used in proper rotation, that is, first in, first out. Solutions are normally stored in type 316 stainless steel, fiberglass-reinforced-plastic (FRP), or plastic-lined tanks.

CONSTRUCTION MATERIALS

Although materials for wastewater neutralization facilities require stability over a wide range of conditions--including exposure to acids, exposure to salts, exposure to alkalies, and possible exposure to solvents--the challenge of the environment is usually limited to a few atmospheres of pressure and temperatures of less than 70° C (158° F).

Traditional materials for wastewater piping, tanks, pumps, and accessories include carbon steel, cast iron, tile, and concrete. These materials provide favorable mechanical properties, practical cost, and adequate inertness for mildly corrosive environments. The traditional materials are of limited applicability, however, to the highly corrosive situations that are characteristic of wastewater neutralization systems. The approach to system design involves the selection of substitute materials that possess adequate inertness or the use of inert coatings to shield the traditional materials from exposure to the corrosive environment. Substantial information has been compiled regarding the corrosion resistance of materials [36,40,41]. In addition, authoritative information is available on the handling of corrosive substances from chemical manufacturers, from vendors of fluids storage and handling equipment, and from vendors of protective coatings.

A tabulated guideline rating of materials for neutralization service is given in Table 6. The tabulation is not all-inclusive and should be considered only for preliminary evaluation because material performance may differ with wastewater characteristics and mode of application. The possible influence of infiltration, thermal shock, and the presence of oils, solvents, and the like also warrants evaluation. Corrosion-resistant materials can be broadly classified as masonry, metals, plastics, and elastomers.

Process Design for Neutralization and Precipitation 93

TABLE 6. GUIDELINE[a] MATERIALS SELECTION CHART FOR WASTEWATER

Chemical	Carbon Steel	316 Stainless	Alloy 20	Duriron™ and Durichlor™	Carbon and Graphite	Butyl Rubber	Polyethylene	Polyvinyl Chloride	Polypropylene	Bisphenol A Polyester	Polyurethanes	Epoxy Polyamide	Fluorocarbon
Acetic acid	N	A	R	R	R	F	R	A	A	F	A	F	R
Aluminum sulfate	N	F	R	R	R	R	R	R	R	R	R	R	R
Ammonium hydroxide	R	R	R	A	R	F	R	R	R	N	F	A	R
Calcium chloride	R	F	A	R	R	R	R	R	R	R	R	R	R
Calcium hydroxide	R	R	R	R	R	R	R	R	R	R	A	R	R
Chromic acid	F	F	A	A	R	N	R	F	F	F	A	R	R
Fatty acids	N	R	R	R	R	N	N	F	N	A	F	N	R
Ferric chloride	N	N	N	A	A	R	R	R	R	R	R	A	R
Ferrous chloride	N	N	F	R	R	F	R	R	R	R	F	R	R
Ferrous sulfate	N	N	R	R	R	R	R	R	R	R	R	R	R
Hydrochloric acid	N	N	A	R	R	A	R	R	R	R	F	A	R
Hydrofluoric acid	F	N	N	A	F	F	R	F	F	A	F	N	R
Nitric acid	N	A	R	R	N	A	N	A	F	A	A	N	R
Phosphoric acid	N	F	R	R	R	A	R	R	R	R	R	A	R
Sodium bisulfite	N	R	R	F	R	R	R	R	R	R	F	R	R
Sodium carbonate	A	A	R	R	R	R	R	R	R	R	R	R	R
Sodium hydroxide	F	A	R	R	R	A	R	R	R	R	F	R	A
Sodium hydrosulfite	F	R	R	F	R	A	R	R	R	R	R	R	R
Sodium hypochlorite	N	F	R	R	N	R	R	R	R	N	A	N	R
Sulfuric acid	F	F	R	R	R	F	R	R	R	F	F	A	R
Sulfurous acid	N	R	R	N	R	F	N	R	R	R	A	A	R
Solvents	R	R	R	R	R	F	N	N	N	R	A	R	R

Note: R is resistant, A is applicable, F is fair, and N is not recommended.

[a] Material selection requires verification for specific situation.

Sources: Adapted from P.N. Cheremisinoff, I. Fideli, and N.P. Cheremisinoff, "Corrosion Resistance of Piping and Construction Materials," *Pollution Engineering*, 54(8):23-26, 1973; J. Javetski, "Solving Corrosion Problems in Air Pollution Control Equipment," *Power*, 122(6):80-87, 1978; R.H. Perry and C.H. Chilton, *Chemical Engineers Handbook* (5th ed.), McGraw-Hill, New York, 1973; Portland Cement Association (PCA), *Effect of Various Substances on Concrete and Protective Treatments, Where Required*, General Information Bulletin 3, PCA, Skokie IL, 1968.

Masonry

Concrete is the traditional masonry material for tanks. The cement binder contains several calcium compounds, including aluminates, silicates, carbonates, and hydroxides. The calcium hydroxide reacts readily with most acids and some salts, the loss of which weakens the concrete by attendant decomposition of aluminates and silicates that can progress to material failure. Corrosion of steel reinforcement can also cause concrete to fail.

The first step in the engineering of concrete tanks for wastewater neutralization is the specification of quality concrete made with portland cement that is low in tricalcium aluminate. It should have a water-to-cement ratio of less than 0.50 and a minimal cement content of 6 bags/yd^3 (8 bags/m^3). The concrete should be consolidated during placement and kept moist at a temperature above 10° C (50° F) for at least a week. The finish should be smooth, clean, and free of voids. The concrete should be well cured (more than 28 d) and dry brush blasting is the recommended surface preparation before application of coatings [41].

Ideally, coatings should be inert to the environment, should form a strong bond with the concrete surface, and should possess sufficient elasticity to adjust to temperature changes and to span cracks. A two-layer coating recommended for the usual service between pH 1 and 12 consists of a 6.3-mm (0.25-in) base surface of glass-reinforced epoxy polyamide covered by a 1.0-mm (0.04-in)--dry fiber thickness--coating of polyurethane elastomer. For service with hydrofluoric acid, synthetic fiber or fabric reinforcement (e.g., Dynel, Dacron) may be used in lieu of fiberglass. Fiberglass is the usual reinforcing material for plastic products, but the use of synthetic fiber reinforcement may be advisable for service with HF or NaOH. References 36 and 42 contain descriptions of other coating systems having application to wastewater neutralization service.

In services where tank interiors are expected to be subjected to physical impact, acid brick linings are superior to resin coatings. The brick may be clay/shale types for general service or carbon/graphite types for service with hydrofluoric acid. The bricks are laid over an appropriate membrane [e.g., 0.64-cm (0.25-in) butyl rubber, polyvinyl chloride, or polyvinylidene chloride] which is then cemented to the concrete. The brick is laid with 8.2-mm (0.32-in) joints of Furan, epoxy, or polyester.

It is important to recognize that lined tanks are subject to damage by hydrostatic pressure if external ground water rises above the liquid level existing in the tank. Concrete tanks may also be subject to damage as a result of exfiltration of corrosives into soil adjacent to the tanks. For these reasons, sound design practice should consider the need for protection of below-ground portions of tanks by external coatings, limestone backfill, and so on.

Metals

Unless used in conjunction with protective coatings, carbon steel and cast iron have poor durability in the presence of the acids and salts usually present in wastewater neutralization systems. Table 6 provides guidance for application of metals.

Stainless steels, such as type 316, are often used in corrosive situations where carbon steel is unsuitable. Stainless steels perform best under oxidizing conditions. They are not recommended under reducing conditions or in the presence of halides. Medium alloys, such as Alloy 20, are also used in corrosive service, particularly with sulfuric acid. The metals have excellent mechanical properties for use in pumps and are immune to solvents normally found in wastewater systems.

High silicon irons, such as Duriron^{TM} and Durichlor^{TM}, are normally applicable to corrosive service except for hydrofluoric acid or strong alkalies. Application is usually limited to pumps and accessories. High alloys and exotic metals are rarely used in wastewater neutralization because more economical alternatives are available for the low-pressure, moderate-temperature conditions.

An assortment of coated steel pipe, pumps, and fittings is available directly from vendors. The lined units combine the inertness of a preferred plastic/elastomer lining with the structural/mechanical integrity of steel.

Linings can be installed in tanks at the fabrication shop or in the field. Surface preparation is extremely important to the performance integrity of lined tanks. White metal blast cleaning is preferred over power tool cleaning, solvent cleaning, or pickling. Blast cleaning is accomplished most practically in shops with automatic equipment. A shop primer of 0.08 mm (3 mils) of ethyl silicate inorganic zinc may be applied to protect the surface pending topcoating. A two-layer topcoat consisting of 3.05 mm (120 mils) of glass-reinforced epoxy polyamide covered by 0.51 mm (20 mils) of polyurethane elastomer is adequate for many situations. Other resin coatings are also popular for corrosive service, as are elastomer linings. Stainless steel substrates are recommended for situations where a significant increase in inertness over carbon steel is required.

Plastics

Particular attention in the selection of materials for wastewaters containing solvents or hydrofluoric acid is warranted. Fluorocarbon plastics such as Teflon^{TM} and Kynar^{TM} are among the most inert materials available for handling corrosives and perform well in wastewaters containing solvents, free halogens, or hydrofluoric acid. Fluorocarbons have the limitation of being relatively expensive and somewhat porous when applied in thin films. Manufacturers of FRP have an assortment of resins and reinforcement fibers available to meet various service specifications. Because glass is attacked by hydrofluoric acid or strong alkali, it is sometimes necessary to use resistant synthetic fibers for reinforcement in these services.

FRP tanks provide excellent corrosion resistance at reasonable cost. The tanks are shop fabricated in sizes of up to about 37,850 l (10,000 gal)--larger sizes are field fabricated. Tanks fabricated from FRP cost 30 percent more than carbon steel tanks and are about half the cost of stainless steel tanks. The installed cost of FRP pipe is similar to carbon steel, although it is limited in strength and span length. Tanks and pipes fabricated from FRP usually use fiberglass reinforcement in conjunction with bisphenol-A fumarate polyester or vinyl ester resins. Epoxy resins or solvent-resistant furan resins can be employed to suit specific situations.

Unreinforced thermoplastic materials are also used for tanks, pipes, pumps, and linings. As a group, thermoplastics such as polyethylene, polypropylene, and polyvinyl chloride provide excellent resistance to dilute acids, alkalies, and salt solutions. In wastewater service they offer selective resistance to reactive chemicals, such as nitric acid and free halogens, and are of limited application in the presence of solvents. Fluorocarbon thermoplastics feature superior corrosion resistance for virtually all services encountered in wastewater neutralization. The use of fluorocarbon thermoplastics is limited by its relatively high cost, weak bonding to substrates, excess porosity of thin films, and narrow usable temperature range.

Elastomers

Plastic coatings and thermoplastic linings have replaced elastomer linings for some uses, but elastomer linings and membranes of natural rubber, butyl, or Hypalon remain in extensive use. Butyl rubber and Hypalon offer good resistance to oxidizing and nonoxidizing acids, good resistance to salt solutions, fair resistance to alkalies, and poor resistance to strong organic solvents. Elastomer formulations are available to suit a wide range of exposure conditions. Gaskets, hoses, and linings of Viton^{TM}, a fluoroelastomer, are used where resistance to solvents is important.

5
Cost Estimation

Cost estimate guidelines are useful for preliminary assessments of alternative wastewater neutralization/precipitation systems, but these systems are unsuited to simplistic cost functions that ignore differences in wastewater characteristics, fluctuations, sludge production, sludge characteristics, and effluent requirements. A more flexible and credible approach to cost estimation involves the accumulation of estimates of system components that are more representative of specific neutralization and precipitation situations. Source information for this "building block" procedure has been published for 30 water treatment unit processes [43].

Definitive cost estimates for engineering projects are prepared only after the completion of most of the engineering effort on the project and the acquisition of extensive contractor and vendor price quotations. The pre-engineering type of estimate presented herein is classified as an order of magnitude estimate with a range of ± 40 percent [36].

Some of the information in Estimating Costs for Water Treatment as a Function of Size and Treatment Efficiency [44] has been adapted to the preparation of guideline estimates for a sample neutralization system, which is described in Example 8 of the appendix. The procedures outlined in the example are amenable to adjustment to various systems incorporating different design concepts, different allowances for backup capability, different construction materials, and different effluent objectives.

Construction, operating, and maintenance cost data shown in this section's illustrations are based on January 1978 dollars. The maintenance costs include the cost of periodic replacement of parts necessary to keep the process operable but do not include the cost of chemical reactants fed to the process. Chemical costs are added separately, and labor costs are taken as $10/h.

The construction costs obtained from the curves are not the final capital cost for the unit process. Allowances must be added to the curve values for costs associated with special site work, general contractor overhead and profit, engineering, land, and legal, administrative, and financing activities. The introduction of these costs into an estimate is illustrated in Example 8 of the appendix.

Estimates of construction costs are intended to be representative of national average price levels as of January 1978. The estimates may be adjusted by application of the Water Quality Office - Sewage Treatment Plant (WQO-STP) Cost Index to specific locations and dates or by more definitive procedures outlined in Reference 43.

MIXED REACTOR COSTS

Construction

Mixed reactor construction costs are estimated in Figure 24 for reinforced concrete tanks ranging in size from 14.2 m^3 (502 ft^3) to 70.8 m^3 (2,500 ft^3). The tanks are coated with fiberglass epoxy and polyurethane for acid resistance. Variable speed, pitched turbine, stainless steel agitators are used that provide a unit power output of 0.288 kW/m^3 (0.036 hp/ft^3).

Operation and Maintenance

Electric power requirements may be estimated by using a 70-percent motor efficiency factor to the output power value. Maintenance costs can be estimated as 0.3 h/d of operating labor plus $80/yr times (tank volume)$^{2/3}$.

SODIUM HYDROXIDE FEED SYSTEM COSTS

Construction

Cost estimates of systems for feeding of 50-percent liquid sodium hydroxide (NaOH), containing 0.76 kg/l (6.36 lb/gal), were adapted from estimates for feed of liquid alum [43]. Fifteen days of storage in coated steel tanks are provided. The tanks are assumed to be located indoors for small installations and outdoors for larger installations. Outdoor tanks are insulated and heating is provided to prevent crystallization.

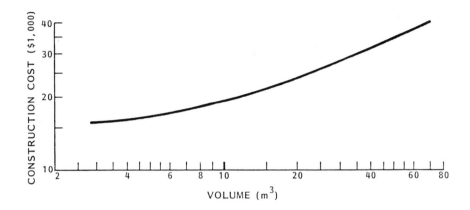

Figure 24. Mixed reactor construction cost.

Dual head metering pumps are used to pump undiluted liquid from storage directly to the point of application. A standby metering pump is included for each installation. The estimates include 37 m (150 ft) of pipe as well as miscellaneous fittings and valves for each metering pump.

Figure 25 presents construction costs for feed of liquid NaOH and an estimated overall cost relation for feed of NaOH with instrumented control of pH. The instrumentation package includes an electrode assembly, signal transmitter, pH recorder-controller, control valves, instrument panel, miscellaneous hardware, and installation.

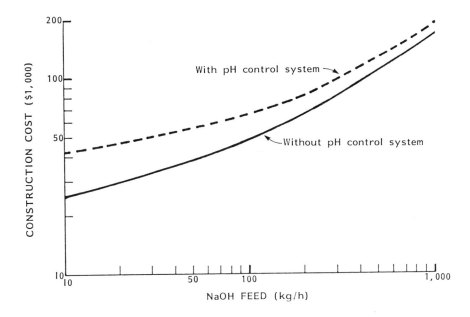

Figure 25. Construction cost of liquid NaOH feed system.

Operation and Maintenance

Operation and maintenance costs are similar to those presented for feed of liquid alum (Figure 26). If a pH control system is provided, a surcharge of $22,000/yr is added to the value obtained from Figure 26.

HYDRATED LIME FEED SYSTEM COSTS

Construction

Cost estimates for hydrated lime feeding facilities were adapted from published estimates for feed of dry alum [43]. For large volume use, the systems employ bulk handling and storage of reagent with pneumatic transfer from delivery trucks. Smaller systems use bag shipments of reagent with bag loaders to reagent feeders. All hopper facilities include dust collectors.

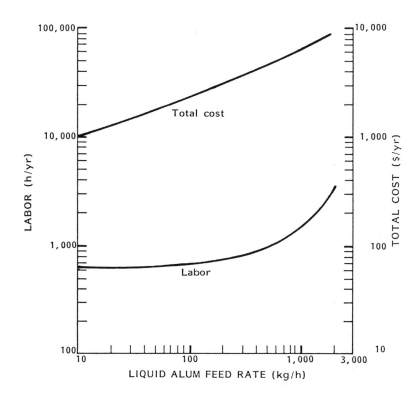

SOURCE: U.S. Environmental Protection Agency, Estimating Costs for Water Treatment as a Function of Size and Treatment Efficiency, EPA 600/2-78-182, NTIS No. Pb 285-274, Aug. 1978.

Figure 26. Operation and maintenance costs of liquid alum feed systems.

Volumetric feeders are installed for low consumption use, and mechanical weight belt feeders are used for large installations. The feeders discharge to slurrying tanks inasmuch as all lime is fed to the process in slurry form.

Construction cost estimates for hydrated lime feed are presented in Figure 27. The figure also presents an estimated overall cost relation for feed of hydrated lime slurry with instrumentation for control of pH. The instrumentation package includes a recycle lime slurry loop, an electrode assembly, a signal transmitter, a pH recorder-controller, control valves, instrument panel, miscellaneous hardware, and installation.

Operation and Maintenance

Operation and maintenance costs for hydrated lime feed systems are 1.2 times the cost given in Figure 28 for feed of dry alum. If a pH control system is provided, a surcharge of $30,000/yr is added to the value obtained above.

Some additional reference material on cost estimation, abstracted from Estimating Costs for Water Treatment as a Function of Size and Treatment Efficiency [43], is presented in the following pages to assist designers with cost estimation. The similarities between components listed and those of neutralization and precipitation systems is sufficient in many cases to enable the calculation of guideline cost estimates by analogy. For instance, the cost information presented for the feeding of anhydrous ammonia has direct application to some neutralization situations and may be adapted to similar feeding operations involving carbon dioxide or sulfur dioxide. The cost information on the feed of liquid alum is analogous to liquid reagent feed operations pertaining to sulfuric acid, sodium hydrosulfide, and so forth.

For dry reagents the cost information on feed of dry alum is analogous to the feed of dry reagents, such as soda ash and sodium bisulfite. Adjustments for different materials of fabrication, different solubilities, different bulk densities, and different solution rates may be required.

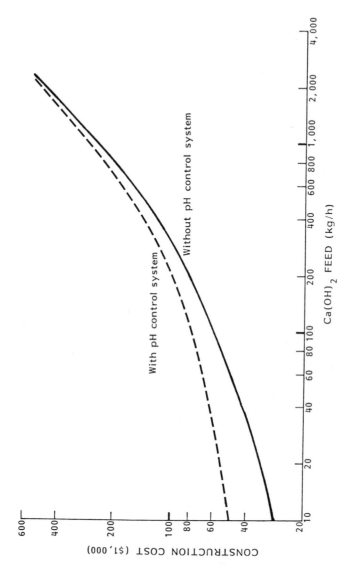

Figure 27. Construction costs of hydrated lime slurry feed system.

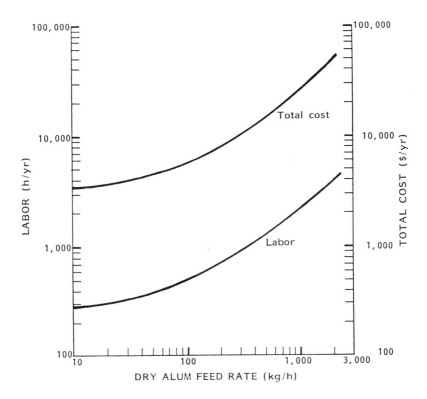

SOURCE: U.S. Environmental Protection Agency, *Estimating Costs for Water Treatment as a Function of Size and Treatment Efficiency*, EPA 600/2-78-182, NTIS No. Pb 285-274, Aug. 1978.

Figure 28. Operation and maintenance costs of dry alum feed systems.

The cost information on polymer feeding and flocculation is directly applicable to neutralization and precipitation technology. The costs for project services (e.g., contractor overhead, legal, and interest) are also representative of neutralization and precipitation systems. More detailed cost information is available in Reference 43.

AMMONIA FEED FACILITIES

Ammonia may be fed in two forms--anhydrous ammonia or aqua ammonia. Anhydrous ammonia is purchased as a pressurized liquid and is converted to the gaseous form in evaporators and ammoniators for feeding to the process. Aqua ammonia is a 29.4-percent solution of ammonia in water. Aqua ammonia is dispensed as a liquid directly to the point of application.

Aqua ammonia is usually available near large cities and is most commonly used in larger plants. A technical disadvantage of anhydrous ammonia can be observed if the gas produced by the ammoniator is used to produce a high strength solution to be fed to the application point. In certain cases, magnesium precipitation occurs as a result of pH elevation caused by ammonia solution. Magnesium precipitation can severely restrict effective ammoniator capacity.

Anhydrous Ammonia Construction Costs

The construction cost curve in Figure 29 includes bulk ammonia storage for all feed rates, with a 10-d storage period. The storage tanks include the tank and its supports, a scale, a dry inert gas padding system, and all required gauges and switches. The ammonia feed system consists of an evaporator for flows in excess of 907 kg/d (2,000 lb/d), an ammoniator, and flow-proportioning equipment. Dry ammonia gas, rather than a high strength ammonia solution, was assumed to be fed directly to the point of application.

Anhydrous Ammonia Operation and Maintenance Costs

Figure 30 presents the cost curves for operation and maintenance of anhydrous ammonia feed facilities. Electrical energy requirements are for heating, lighting, ventilating of the ammoniator building, and operation of the evaporators. Evaporators are only included for systems of 907 kg/d (2,000 lb/d) or greater, and evaporator energy requirements were calculated on the basis of 23.8 kWh/metric ton (81,205 Btu/short ton) of ammonia.

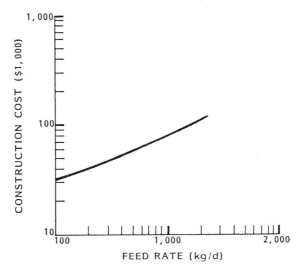

SOURCE: U.S. Environmental Protection Agency, Estimating Costs for Water Treatment as a Function of Size and Treatment Efficiency, EPA 600/2-78-182, NTIS No. Pb 285-274, Aug. 1978.

Figure 29. Construction cost for anhydrous ammonia feed facilities.

Maintenance materials were based on operating experience for similar size chlorination facilities. Anhydrous ammonia costs are not included in the maintenance material costs. Labor requirements are for transfer of the bulk anhydrous ammonia from the delivery truck or rail car to the onsite ammonia storage tank, plus daily operation and maintenance requirements. A bulk unloading time of 3 h per shipment was used. Operation and maintenance requirements varied from 1.5 h/d for the smaller systems to 3 h/d for larger systems.

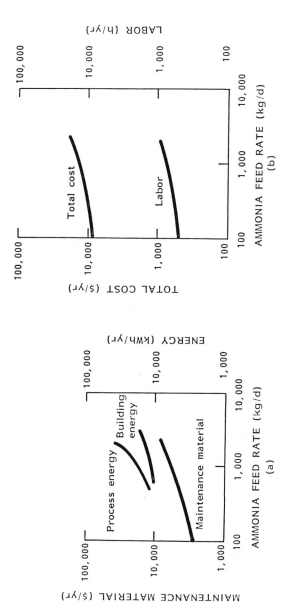

Figure 30. Operation and maintenance costs for anhydrous ammonia feed facilities: (a) maintenance material and energy and (b) total cost and labor.

SOURCE: U.S. Environmental Protection Agency, Estimating Costs for Water Treatment as a Function of Size and Treatment Efficiency, EPA 600/2-78-182, NTIS No. Pb 285-274, Aug. 1978.

ALUM FEED SYSTEM COSTS

Construction

Construction costs for liquid and dry alum feed systems are shown in Figure 31.

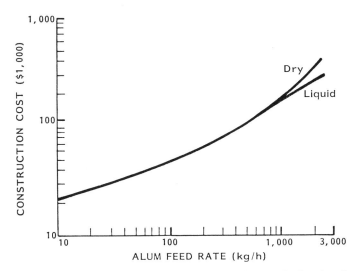

SOURCE: U.S. Environmental Protection Aency, <u>Estimating Costs for Water Treatment as a Function of Size and Treatment Efficiency,</u> EPA 600/2-78-182, NTIS No. Pb 285-274, Aug. 1978.

Figure 31. Construction cost for alum feed systems.

Liquid Alum

The cost data for liquid alum feed systems are based on a density of 1.2 kg/l (10 lb/gal). Fifteen days of storage are provided using FRP tanks. The FRP tanks are assumed to be uncovered, located indoors for smaller installations and outdoors for larger installations. Outdoor tanks are covered and vented, with insulation and heating provided to prevent crystallization, which occurs at temperatures below -1.1° C (30° F).

Dual head metering pumps are used to pump liquid alum from the storage tank and dispense the flow directly to the point of application. No provision is made for dilution of the liquid alum before application. A standby metering pump is included for each installation. All pipe used to convey the liquid alum is type 316 stainless steel; 37 m (150 ft) of pipe, along with miscellaneous fittings and valves, are included for each metering pump.

Dry Alum

The cost data in Figure 31 are based on use of commercial dry alum with a density of 961 kg/m^3 (60 lb/ft^3). A 5-min detention period is required in the dissolving tank, and water is used at a rate of 16.7 l/kg (2 gal/lb) of alum. Fifteen days of dry alum storage in mild steel storage hoppers located indoors are assumed. Pneumatic conveyance of alum from bulk delivery to the hoppers is used, with the blower located on the delivery truck. The maximal hopper size is 170 m^3 (6,000 ft^3). For installations too small for bulk delivery, bag loaders are used on the feeder. All hopper facilities include dust collectors.

Volumetric feeders for the smaller installations and mechanical weigh belt feeders for the larger installations and their respective solution tanks are located directly beneath the storage hoppers, eliminating the need for bucket elevators or other conveyance devices from underground storage. Such installation does, however, make the building cost somewhat greater than other possible arrangements. Conveyance from the solution tanks to the point of application is by dual head diaphragm metering pumps.

Operation and Maintenance

Figures 32 and 33 present operation and maintenance costs for both liquid and dry alum feed systems. Electrical requirements are for solution mixers, feeder operation, building lighting, ventilation, heating, and, for larger liquid feed installations, heating of outdoor storage tanks. The sharp decrease in the building energy curve for high feed rates is attributable to

Cost Estimation 111

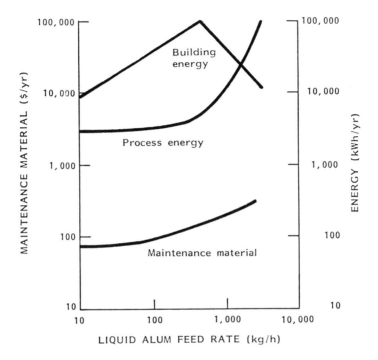

SOURCE: U.S. Environmental Protection Agency, <u>Estimating Costs for Water Treatment as a Function of Size and Treatment Efficiency</u>, EPA 600/2-78-182, NTIS No. Pb 285-274, Aug. 1978.

Figure 32. Operation and maintenance costs for liquid alum feed systems.

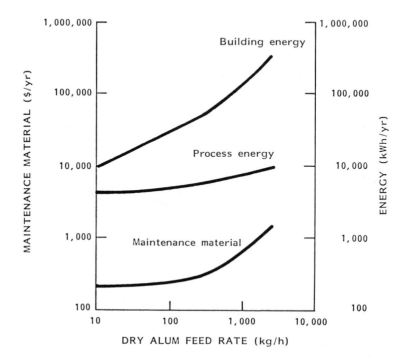

SOURCE: U.S. Environmental Protection Agency, <u>Estimating Costs for Water Treatment as a Function of Size and Treatment Efficiency</u>, EPA 600/2-78-182, NTIS No. Pb 285-274, Aug. 1978.

Figure 33. Operation and maintenance costs for dry alum feed systems.

the use of outdoor storage tanks, as contrasted with the use of indoor storage tanks at the lower flow rates.

Maintenance material costs were estimated to be 3 percent of the manufactured equipment cost, excluding storage tank cost. Alum costs are not included in the maintenance material costs.

Labor requirements consist of time for chemical unloading and routine operation and maintenance of feeding equipment. Liquid and dry alum unloading requirements were calculated at 1.5 h per bulk truck delivery and 5 h per 22,680 kg (50,000 lb) respectively. For dry feed installations using alum from bags, 8 h were used per 7,258 kg (16,000 lb) removed and fed to the bag loader hopper. Time for routine inspection and adjustment of feeders is 10 min per feeder per shift for dry feed and 15 min per metering pump per shift for liquid feed. Maintenance requirements are 8 h/d for liquid metering pumps and 24 h/d for solid feeders and the solution tank.

POLYMER FEED SYSTEM COSTS

Construction

Cost estimates for polymer feed systems shown in Figure 34 are based on the use of dry polymers, fed manually to a storage hopper located on the chemical feeder. Chemical feed equipment is based on preparation of a 0.25-percent stock solution. No provision has been made for standby or redundant equipment because polymer normally would be used only as a coagulant aid or a filter aid; therefore, an equipment breakdown could be tolerated for a short period of time while equipment was being repaired.

Costs are based on a system consisting of a feeder, a solution tank, the water piping to the feeder, a polymer solution line out of the building, installation labor, the cost of housing for the feeder/mixer, and a bag storage area for up to 15 d of storage.

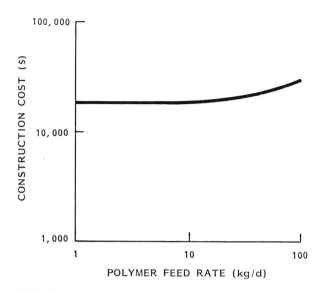

SOURCE: U.S. Environmental Protection Agency, Estimating Costs for Water Treatment as a Function of Size and Treatment Efficiency, EPA 600/2-78-182, NTIS No. Pb 285-274, Aug. 1978.

Figure 34. Construction cost of polymer feed systems.

Operation and Maintenance

Figure 35 presents the estimated operation and maintenance costs for feeding of a 0.25-percent polymer solution. Energy requirements for the feeder and metering pump were calculated using motor horsepower requirements recommended by manufacturers. Building energy requirements are based on completely housed systems.

Maintenance material costs used are 3 percent of manufactured equipment and pipe/valve costs. These costs do not include the cost of polymer.

Cost Estimation   115

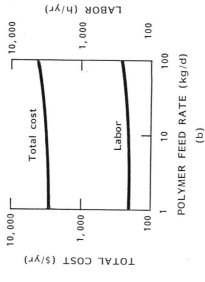

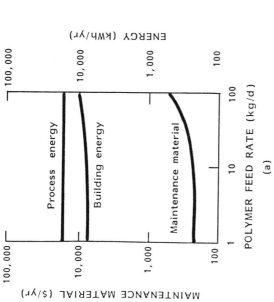

SOURCE: U.S. Environmental Protection Agency, Estimating Costs for Water Treatment as a Function of Size and Treatment Efficiency, EPA 600/2-78-182, NTIS No. Pb 285-274, Aug. 1978.

Figure 35. Operation and maintenance costs for polymer feed systems: (a) maintenance material and energy and (b) total cost and labor.

Labor requirements are for bag unloading (1 h per ton of bags), the dry chemical feeder (110 h/yr for routine operation and 24 h/yr for maintenance), and the solution metering pump (55 h/yr for routine operation and 8 h/yr for maintenance).

FLOCCULATION SYSTEM COSTS

Construction

Estimated costs for vertical turbine flocculators are given in Figure 36. Estimated flocculation basin costs are for 3.7-m- (12-ft-) deep, rectangular, reinforced concrete structures, with common wall construction where the total basin volume exceeds 354 m^3 (12,500 ft^3). A length-to-width ratio of approximately 4:1 is used for basin sizing, and the maximal basin size is 354 m^3 (12,500 ft^3). Structural costs for vertical turbine flocculators are somewhat higher than for the horizontal paddle type because of required structural support above the basin. Costs were calculated for vertical flocculators with total basin volumes between 51 and 708 m^3 (1,800 and 25,000 ft^3).

G values of 20, 50, and 80 s^{-1} were used to calculate manufactured equipment costs. All drive units are variable speed to allow maximal flexibility. Although common drive for two or more parallel basins is commonly used, the estimated costs were calculated using individual drive for each basin.

Operation and Maintenance

Figure 37 presents operation and maintenance costs for G values of 20, 50 and 80 s^{-1} for vertical turbine flocculators. Energy requirements for G values of 20, 50, and 80 s^{-1} were based on the respective energy per unit volume requirements of 0.0005, 0.0029, and 0.0081 kW/m^3 (0.19, 1.1, and 3.07 x 10^{-4} hp/ft^3). An overall motor efficiency of 60 percent is used.

Maintenance material costs are based on 3 percent of the manufactured equipment costs. Although equipment costs vary somewhat with the maximal design value of G, the maintenance material costs are based on a G value of 80 s^{-1}.

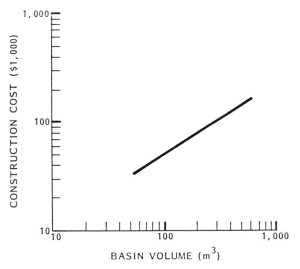

Note.—Agitation velocity gradient, G, = 20, 50, or 80 s^{-1}.

SOURCE: U.S. Environmental Protection Agency, Estimating Costs for Water Treatment as a Function of Size and Treatment Efficiency, EPA 600/2-78-182, NTIS No. Pb 285-274, Aug. 1978.

Figure 36. Construction cost of flocculation for vertical turbine system.

Labor requirements are based on routine operation and maintenance time of 15 min/d per basin [maximal basin volume = 354 m^3 (12,500 ft^3)] and an oil change every 6 mo requiring 4 h per change. No allowance is included for jar test time.

OTHER COSTS

Figures 38 through 40 provide cost estimates for general contractor overhead and fees, legal, and administrative costs, and the cost of interest.

118 Removal of Metals from Wastewater

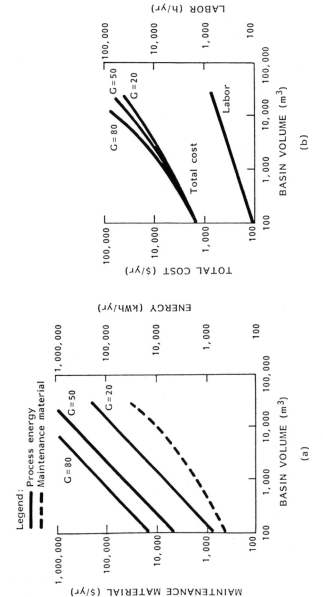

Note.--G refers to the agitation velocity gradient, which is measured in reciprocal seconds (s^{-1}).

SOURCE: U.S. Environmental Protection Agency, Estimating Costs for Water Treatment as a Function of Size and Treatment Efficiency, EPA 600/2-78-182, NTIS No. Pb 285-274, Aug. 1978.

Figure 37. Operation and maintenance costs for flocculation systems: (a) maintenance material and process energy and (b) total cost and labor.

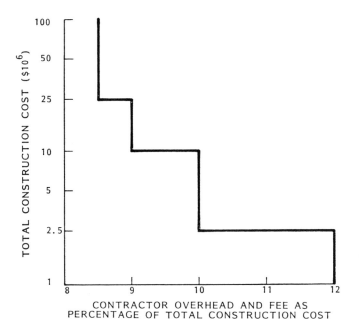

SOURCE: U.S. Environmental Protection Agency, Estimating Costs for Water Treatment as a Function of Size and Treatment Efficiency, EPA 600/2-78-182, NTIS No. PB 285-274, Aug. 1978.

Figure 38. General contractor overhead and fee as percentage versus total construction cost.

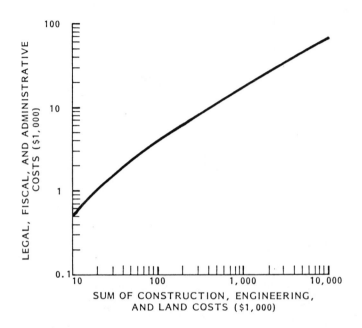

SOURCE: U.S. Environmental Protection Agency, Estimating Costs for Water Treatment as a Function of Size and Treatment Efficiency, EPA 600/2-78-182, NTIS No. Pb 285-274, Aug. 1978.

Figure 39. Legal, fiscal, and administrative costs for projects less than $10 million.

Cost Estimation   121

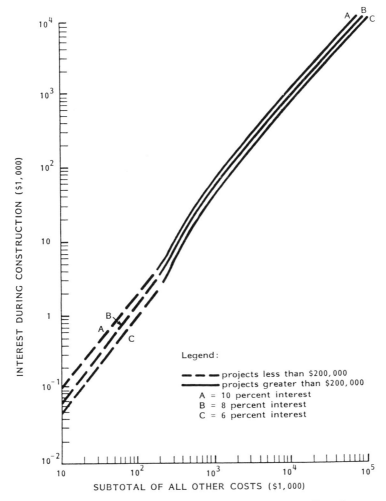

SOURCE: U.S. Environmental Protection Agency, Estimating Costs for Water Treatment as a Function of Size and Treatment Efficiency, EPA 600/2-78-182, NTIS No. Pb 285-274, Aug. 1978.

Figure 40. Interest during construction for projects less than and greater than $200,000.

6
Suspended Solids Separation

The treatment of municipal as well as industrial wastewaters usually results in the formation of slurries that are high in suspended solids. These slurries or sludges are produced either by the concentration of solids originally in the wastewater (such as raw primary sludge in domestic wastewater treatment plants) or new suspended solids formed as a result of the removal process (such as waste-activated sludge or metal hydroxide sludges in plating waste treatment).

In most cases, the only economically feasible treatment for chemical sludges before disposal is the removal of some or most of the water. In this report, the various processes for thickening and dewatering are referred to as solid/liquid separation.

Following coagulation and flocculation, suspended solids are physically separated from the waste stream using such techniques as sedimentation, flotation, centrifugation, or filtration. Perfect solid/liquid separation (clear liquid and bone-dry solid) cannot be achieved by any one method; a combination of devices is necessary to achieve the desired degree of separation economically. The most common approach follows:

- Make an initial split between solids and liquids, using either sedimentation or flotation.
- Use a polishing technique on the separated liquid stream to yield a liquid effluent of high clarity.
- Use a dewatering technique to reduce the moisture content of the separated solids effluent stream.

In selecting an appropriate separation technique, consideration should be given to such factors as:

- current industry practice for the same or similar wastes
- ability to handle the volume and type of waste
- ability to produce an effluent with the desired characteristics
- characteristics of the concentrated solids that are produced
- mechanical simplicity
- reliability of equipment operation
- controllability of the process
- space requirements
- capital and operating costs.

Because the concentration of sludge solids involves the removal of some of the water, the sludge volume may be reduced significantly. For example, the concentration of a 1 percent sludge to a 2 percent sludge results in a 50 percent volume reduction (Figure 41). Transportation in sludge management can be a major expense, so even seemingly small increases in the sludge solids fraction can yield significant savings.

Sludge contains water that is associated with the solids in several ways. Although no universal agreement or terminology exists, the types of water in sludge can be classified as free water, floc water, particle water, and chemically bound water.

Solid particles can settle through free water, which is defined as water that can be removed from sludge by simple gravitational means (thickening). Floc water is trapped between the sludge particles and cannot be removed by gravity alone. The floc particles must be compressed by mechanical means, such as filtration or centrifugation. Because particle water is part of the surface structure of the floc particle, it can be removed only by deforming or breaking the particle. Chemically bound water, the water of hydration, is removed by chemical or thermal means.

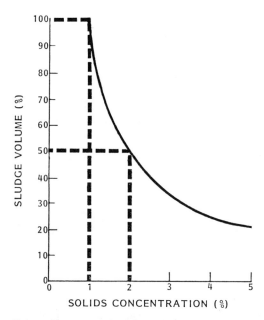

Note.—For every doubling of the solids concentration (e.g., from 1 percent to 2 percent), the volume is reduced by 50 percent.

Figure 41. Sludge volume reduction with the removal of water.

It is sometimes advantageous to know what type of water is to be removed. For example, if a sludge contains a high percentage of chemically bound water, it may be necessary to remove water by thermal means. On the other hand, a large free water component (e.g., a clay slurry) would require only thickening. Thus, the selection of the proper equipment for removing water from sludge requires a knowledge of how the water is distributed in the sludge.

Solid/liquid separation techniques are based on the following three principles:

- The solids are more (or less) dense than the liquid.
- The solids are larger in physical size than the liquid.
- The solids will not volatilize when then liquid is evaporated.

These three principles are illustrated in Figure 42.

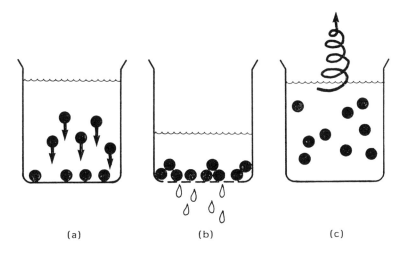

Figure 42. Principles of solid/liquid separation: (a) solids are denser, (b) solids are larger, and (c) solids will not volatilize.

Table 7 lists a number of commercially available solid/liquid separation techniques and the basic principles they employ. Not all of the methods listed are applicable to separating solids from liquids in chemical slurries. Only the most commonly employed techniques are discussed in this section.

TABLE 7. TECHNIQUES FOR ACHIEVING SOLID/LIQUID SEPARATION FOR WASTEWATER SLUDGES

Technique and equipment	Basic principle used[a]
Sedimentation:	
Thickening tank	1
Lamella or tube thickener	1
Lagoon	1
Flotation:	
Air	1
Dissolved air	1
Vacuum	1
Electroflotation	1
Centrifugation:	
Basket	1
Disc	1
Solid bowl	1
Perforated bowl	1 and 2
Hydrocyclone	1
Filtration:	
Vacuum	2
Filter press	2
Belt	2
Gravity	2
Drying:	
Rotary	3
Flash	3
Sand bed	2 and 3

[a] Numbers in this column correspond to the following three principles: 1--the solids are more (or less) dense than the liquid; 2--the solids are larger in physical size than the liquid; and 3--the solids will not volatilize when the liquid is evaporated.

SEDIMENTATION

Sedimentation is the gravity settling of suspended solids from a liquid. Because the liquid phase is clarified as the solids settle out, the process is frequently referred to as clarification and the equipment is called a clarifier. Figure 43 illustrates the position of clarification in a typical wastewater treatment system.

The influent to the clarifier is usually received directly from the flocculation step. The overflow stream from the clarifier is a clear effluent that may be discharged directly to the environment (if the effluent meets legal or self-imposed limitations), undergo further treatment before discharge, or be reused within the process. The bottoms discharge, or sludge, from the clarifier usually has a solids content of no more than 5 percent. To reduce the sludge volume, the bottoms from the clarifier may be fed to a thickener to increase the solids concentration to 5 to 10 percent [44].

A major objective in designing sedimentation hardware is to create a quiescent state in which the suspended particles that have a specific gravity higher than the liquid will settle out. Settling basins must separate a wide variety of suspended solids.

Classes of Particle Settling

The following four classes, or zones, of particle settling are based on the concentration and the tendency of the particles to interact:

- Class 1--free settling
- Class 2--hindered settling
- Class 3--zone settling
- Class 4--compression settling

A particle normally will pass through all four zones during clarification and thickening. Therefore, the requirements for each zone must be considered in the hardware design. The concentration of solids increases as the particle sinks to a lower zone. It is this relationship between time and concentration on which the design of sedimentation equipment is based [44].

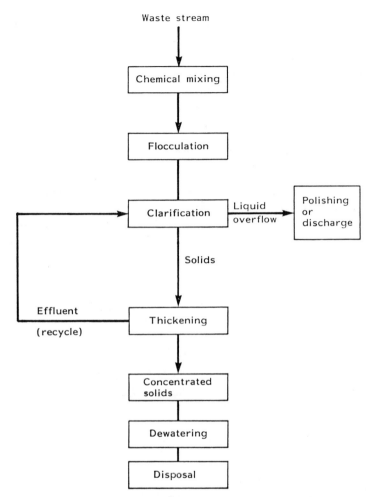

SOURCE: U.S. Environmental Protection Agency, _Controlling Pollution From the Manufacturing and Coating of Metal Products_, EPA 525/3-77-009, 1977.

Figure 43. Typical wastewater treatment system.

Class 1--Free Settling

Class 1 settling refers to the sedimentation of discrete particles in a dilute suspension (e.g., a grit chamber). The particles settle independently without significant interaction with neighboring particles. Settling occurs at a rate directly related to particle size and density and fluid density [44,45]. When particles in suspension reach the settling zone of a clarifier, their settling velocity (V_s) and direction can be altered by the horizontal velocity (V_h) of the liquid passing through the settling zone. If the resulting trajectory takes a particle to the bottom of the sedimentation basin before reaching the outlet zone, it will be removed from the effluent. Referring to Figure 44, a particle must settle through the distance, H, at the velocity, V_s, in the same time (or less) that the liquid is in the basin (i.e., residence time) [44,46].

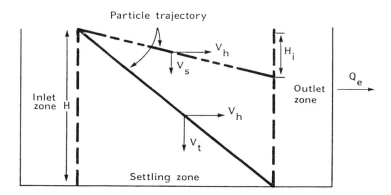

Legend:

Q_i = influent flow rate
H = basin height
V_s = settling velocity
V_h = horizontal velocity

V_t = terminal velocity
Q_e = effluent flow rate
H_i = difference between baffle height and total basin height

Figure 44. Free settling in an ideal sedimentation basin.

In designing a sedimentation basin, a terminal velocity, V_t, is selected, and all particles with settling velocities greater than V_t will be removed from the wastewater. By definition, the terminal velocity is equal to the overflow rate or surface hydraulic loading:

$$V_t = Q/A = \text{overflow rate} \qquad (37)$$

where Q = flow (m^3/d)
A = horizontal area of the sedimentation unit (m^2)

Thus, in Class 1 settling, the flow capacity is independent of the basin height [45]. The basin height, H, however, is related to the design velocity by the equation

$$V_t = H/t \qquad (38)$$

where t = time

If the settling velocity distribution of the particles in the waste streams is determined, the overall particle removal efficiency can be calculated for a given wastewater flow rate and basin size, as explained in Example 9 of the appendix.

In actual practice, free settling without particle interaction is rare. Furthermore, settling never occurs under perfectly quiescent conditions, and the design of the system must take into account many factors. Because it is difficult to relate these factors to laboratory procedures, they are discussed in later subsections (Design; Scale-Up).

Class 2--Hindered Settling

Class 2 settling refers to particles encountering other particles in the suspension, thereby decreasing their settling rate. The extent to which particle interaction occurs depends on the opportunity for particle contact, which is determined by depth of the basin, overflow rate, velocity gradients in the system, and particle concentration and range of sizes [44,45].

Because no mathematical models exist that approximate the settling characteristics of a suspension of flocculant particles, laboratory settling tests are typically performed in columns of the same height as those envisioned for the equipment [i.e., about 3 m (10 ft)]. Satisfactory results can be obtained with a 15-cm- (6-in-) diameter plastic tube, although the column can be of any diameter. Samples usually are collected at 0.6-m (2-ft) intervals, and the solids percentages are determined at each port location (Figure 45). Care should be taken in introducing the suspension into the column to achieve a uniform distribution of particle sizes throughout the test apparatus. Furthermore, convection currents should be avoided by maintaining a uniform temperature throughout the test [44,45].

For each sample analyzed, the percentage of suspended solids removed is plotted on a graph of time versus depth. Between the plotted points, curves of equal percentage removal are drawn. A completed plot, as shown in Figure 45, gives the results of a settling column sedimentation test. The plotted points representing the individual samples have been omitted from the figure. The overall removal of solids for quiescent settling can be calculated for a given solids detention time and depth. For example, for a detention time of t_2 and a depth of h_5, the percent removal may be calculated by:

$$\% \text{ removal} = \sum_{n=1}^{4} \left[(\Delta h_n / h_5) \times (R_n + R_{n+1})/2 \right]$$

$$= [(\Delta h_1/h_5) \times (R_1 + R_2)/2]$$

$$+ [(\Delta h_2/h_5) \times (R_2 + R_3)/2]$$

$$+ [(\Delta h_3/h_5) \times (R_3 + R_4)/2]$$

$$+ [(\Delta h_4/h_5) \times (R_4 + R_5)/2] \tag{39}$$

where h = depth of sediment
R = removal efficiency

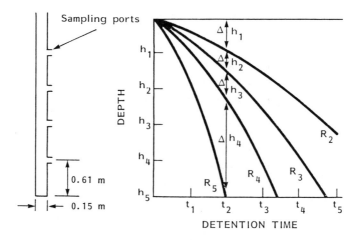

Legend:

R_2 = 80-percent removal efficiency
R_3 = 70-percent removal efficiency
R_4 = 60-percent removal efficiency
R_5 = 50-percent removal efficiency

SOURCE: Metcalf and Eddy, Inc., Wastewater Engineering: Treatment, Collection, and Disposal, McGraw-Hill, New York, 1972.

Figure 45. Settling column and settling curves for flocculent particles.

Given the desired concentration for the overflow, it is possible to determine the sedimentation basin area, residence time, and depth for free settling [44,45].

Classes 3 and 4--Zone Settling and Compression Settling

Whereas the first two settling classifications are normally used to design clarifiers, Classes 3 and 4 are used in designing thickeners for the concentration of solids. Zone settling (Class 3) occurs when the solids that have coagulated in the hindered zone move as a consolidated mass, thereby forming a distinct solid-liquid boundary at the top of the sludge mass. The particle concentration is of an intermediate density, significantly greater than for Class 1 or Class 2 suspensions but less than for Class 4. The resulting interparticle forces cause all the particles to settle effectively at the same rate, as a "blanket," thereby producing a region of clarified liquid.

As the sludge mass settles and the suspended solids are compacted, a compression layer begins to form on the bottom of the settling basin or column. Compression settling (Class 4) is a function of the retention time and the compression depth of both the solid layer and the liquid above it. The rate of compression is very slow compared with the other sedimentation processes [44,45].

To evaluate the zone settling and compression characteristics of a suspension, the designer performs laboratory settling tests similar to those discussed above for sizing free-settling sedimentation basins. Reference 47 describes a method to determine the area requirement for zone settling; if a column of height H_o is filled with a suspension of uniform concentration C_o, the height of the liquid-solid interface will decrease at a rate equal to the slope of the curve in Figure 46. In the design of thickeners, the objective is to maximize the concentration of the sludge. The area required for sludge thickening is determined from the following equation:

$$A = Qt_u/H_o \tag{40}$$

where A = area required for sludge thickening (m^2)
 Q = flow rate into tank (m^3/s)
 H_o = initial height of interface in column (m)
 t_u = time to reach desired underflow concentration, C_u (s)

To calculate t_u, it is necessary to determine the critical concentration, C_2, that controls the sludge-handling capability of the tank. By definition, C_2 is the concentration at the midpoint of the region between hindered settling

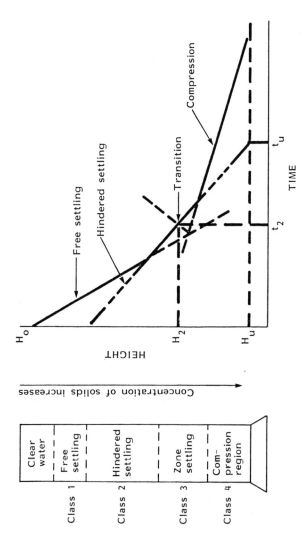

Figure 46. Typical sedimentation curve showing the settling zones and illustrating the graphical calculation of the time required to reach the desired underflow concentration in a thickener.

and compression (i.e., at H_2 in Figure 46). This point may be obtained by extending the tangents to the free-settling and compression regions of the (interface) subsidence curve to the point of intersection and bisecting the resulting angle, as illustrated in Figure 46. Next, the depth, H_u, at which all the solids are at the desired underflow concentration may be calculated from the relationship:

$$H_u = C_o H_o / C_u \qquad (41)$$

where C_o = initial solids concentration (mg/l)
C_u = desired underflow concentration (mg/l)
H_o = initial height of the interface (m)

A horizontal line is constructed at the calculated H_u. The intersection of the tangent at H_2, t_2 and the horizontal line at H_u determines t_u, the time required to reach the desired underflow concentration.

Because thickening is usually accomplished in the same sedimentation basin as clarification, the design of the basin is based on the equation that has the larger area requirement. The sedimentation basin must also provide space on the bottom for storage of sludge during compaction. The amount of space required depends on the rate of compaction of the sludge and the desired residence time. The rate of compaction of settled solids in the compression region may also be determined by settling tests. This rate, which is slow compared with other sedimentation processes, has been found to be proportional to the difference between the depth at a time t and the depth to which the sludge will settle after a long period of time (e.g., 24 h). This can be represented by:

$$H_t - H_\infty = (H_2 - H_\infty) \exp^{-i(t-t_2)} \qquad (42)$$

where H_t = sludge height at the desired residence time, t (m)
H_∞ = sludge depth after an extended time period (m)
H_2 = sludge height at a time, t_2 (m)
i = a constant for a given suspension

Aids to Settling

Coagulants, or coagulant aids, may be added to the wastewater to enhance the coagulation process. Coagulants are water-soluble compounds that increase the electrolyte concentration in wastewater, thereby compressing the electrical double-layer surrounding each particle in suspension. Repulsive forces between particles are reduced and the suspension is destabilized. Iron, aluminum, and calcium salts are among the most effective coagulants. Agglomeration of coagulated solids is promoted largely by interparticle bridging. The type and quantity of coagulant or coagulant aids to be used depend on the nature of the wastewater, that is, the signs, magnitudes, and distribution of surface charge [48].

Design

The design of sedimentation equipment involves careful consideration of the basic parameters that affect sedimentation efficiencies in actual practice. Such factors include the effects of short circuiting, inlet and outlet turbulence, scour velocity, hydraulic loading, and sludge capacity. In the actual design, the range of scale-up factors should reflect the influence of flow variation, wind, and temperature gradients, as well as differences in basic design features, such as tank shape [44,45,49,50].

Scale-up

Hazen [51] showed that, under the following conditions, performance is a function of surface loading alone:

- under quiescent or nonturbulent flow
- when distribution of velocity is uniform over all sections of the tank normal to the general flow direction
- when discrete noninteracting particles exist
- when there is no resuspension of settled particles after they reach the bottom of the tank.

In actual basins, conditions differ in many respects from those in Hazen's original analysis. The most significant of these departures are as follows [49]:

- Currents induced by inlets, outlets, wind, and density differences may cause short circuiting or dead spaces within the tank.
- Turbulence caused by forward velocity or currents in the tank retards settling.
- Flocculent solids may agglomerate into larger particles during passage through the basin.
- Sludge may be scoured and resuspended at high forward velocities.
- When influent solids concentrations are high, particles settle as a mass rather than discretely.

The following subsections provide insight in evaluating various design factors and using settling tests for sizing full-scale equipment. Metcalf and Eddy, Inc. [45], recommends that the value for the design settling velocity or overflow rate determined from the laboratory tests be multiplied by a factor of 0.65, and the detention times be multiplied by a factor of 1.75 or 2.0. These factors are not intended to cover extreme variations in flows or solids loading or to allow for operation at temperatures significantly different from those under test conditions. Furthermore, standby capacity, required for units critical to overall plant performance, should be designed for independence of any given design factors [49].

Short Circuiting

EPA's Process Design Manual for Suspended Solids Removal [49] states that "short-circuiting is accentuated by high inlet velocities, high outlet weir rates, close placement of inlets and outlets, exposure of tank surface to strong winds, uneven heating of tank contents by sunlight, and density differences between inflow and tank contents." Short circuiting is minimized in narrow, rectangular, horizontal-flow tanks and becomes a more serious problem in circular horizontal flow tanks. The degree of short circuiting in circular units can vary widely, depending on the inlet configuration. The best approach to this design problem is to minimize those factors that contribute to short circuiting [48,49,52].

Turbulence

Turbulence caused by drag from net forward velocity may be compensated for by increasing the tank area. Such required increases would vary directly with the forward velocity in the tank and with the desired removal rate [48]. Other sources of turbulence, such as inlet, outlet, wind, and density currents, are largely unpredictable and may increase short circuiting. Furthermore, agglomeration induced by turbulence can alter particle sizes and localized settling velocities [49].

Bottom Scour Velocity

Horizontal (forward) velocities should not exceed the critical velocity at which settled particles are scoured from the bottom of the basin. Camp [53] developed the following definition for the critical velocity from data developed by Metcalf and Eddy, Inc. [45]:

$$V_H = \frac{8k(s-1)g(d)}{f}^{0.5} \tag{43}$$

where V_H = horizontal velocity that will just produce scour
   s = specific gravity of the particles
   d = diameter of the particles
   k = a constant that varies with the type of material being scoured
   f = Darcy-Weisbach friction factor, which depends on the characteristics of the surface over which the flow is taking place and the Reynolds number
   g = gravitational constant

Inlet Design

The inlet to a sedimentation tank must allow for distribution of the incoming flow across the entire cross-sectional area of the tank, while minimizing short circuiting and turbulence factors. As a result, inlet velocities should be kept as low as possible [48,54]. Circular settling basin inlets are either central, in which case the influent is piped to the center of the basin through a conduit (Figure 47a), or peripheral (Figure 47b and c). Inlet design for rectangular tanks, where the distance from inlet to outlet is large, is less critical than for circular tanks.

Design of the inlet well (or feed well) should ensure dissipation of the inlet energy and uniform distribution of the outflow in all radial directions. As a rule, the maximal inlet velocity to a center inlet well should not exceed 0.9 m/s (3.0 ft/s), and the outflow velocity should not exceed 4.6 m/min (15.0 ft/min) [48].

Peripheral feed units typically distribute the influent through orifices located in an influent raceway on the entire periphery of the settling basin. Examples of peripheral feed units are shown in Figure 47b and c, in which inlet ports discharge outside a deep peripheral baffle and flow passes under this baffle to enter the tank. Other units are commercially available in which the inlet line to the tank discharges tangentially into a tapered race located behind a similar skirt baffle [49]. In model studies, this type of peripheral unit showed significantly higher removals of iron floc than a similarly loaded center feed unit [55]. This effect was attributed to more favorable conditions for particle agglomeration in the peripheral feed model [49].

In the design of rectangular tanks, the influent should be distributed uniformly over the basin cross section to avoid density currents along the bottom. Inlet ports are normally designed for velocities of 4.6 to 9.1 m/min (15.0 to 30.0 ft/min). Baffles should be placed several inches below the water surface and should extend to a point 15 to 30 cm (0.5 to 1.0 ft) below the inlet point to aid in distributing the water uniformly over the basin cross section [48].

Basin Depth

Although the basin depth does not directly affect the sedimentation process, consideration should be given to proper depth for storage of settled solids, prevention of scour velocity, and prevention of updrafts that may disturb previously settled solids. Basin depths usually vary from 2.0 to

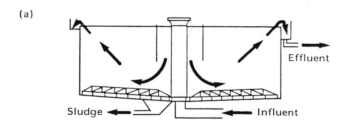

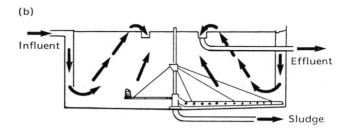

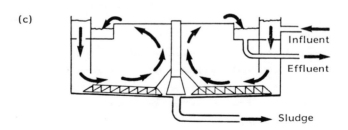

SOURCE: U.S. Environmental Protection Agency, Process Design Manual for Suspended Solids Removal, EPA 625/1-75-003a, NTIS No. Pb 259-147, Jan. 1975.

Figure 47. Typical circular clarifier configurations: (a) center feed clarifier with scraper sludge removal system, (b) rim feed--center takeoff clarifier with hydraulic suction sludge removal system, and (c) rim feed--rim takeoff clarifier.

4.6 m (6.6 to 15.0 ft), with depths of 2.4 to 3.7 m (8.0 to 12.0 ft) being the most common [48].

Outlet Design

The overflow weir should be designed so that the clarified effluent can be removed from the tank without causing localized high velocity updrafts. For circular basins, undesirable density currents will be mitigated by a peripheral weir and by a basin depth that increases with diameter. Small rectangular basins require that a launder be placed along the end wall of the basin. Larger basins should have two or more lateral launders across the basin to prevent localized updrafts [48].

Equipment and Operation

Settling tanks should include the following [44]:

- an inlet zone that distributes the suspension over the cross-sectional area of the basin
- an effective settling zone, where the majority of the settling occurs
- a solids removal or sludge zone, where the solids are stored
- an outlet zone, where the clarified wastewater is collected and discharged.

Thus, the sedimentation tank serves to remove settleable solids and produce a clarified effluent, collect and discharge an underflow stream or sludge, and, where possible, act as a sludge thickener [50]. Table 8 shows the approximate loadings for some common suspensions. In practice, however, the design of sedimentation equipment should be based on laboratory settling tests of representative feed samples wherever possible. Specific information on equipment sizing may be obtained from vendors. Hardware is available that combines chemical (flash) mixing, flocculation, clarification, and sludge thickening in an integral unit.

TABLE 8. LOADINGS FOR SOME COMMON SUSPENSIONS

Types of suspension	Settling velocity (cm/s)	Surface loading (1/h-m^2)	Detention period[a] (h)
Aluminum and iron floc	0.083	1,020-3,050	2-8
Calcium carbonate precipitates	0.042	1,020-4,900	1-4

[a] For a 3-m-deep tank.

SOURCE: R. P. Canale and J. A. Borchardt, "Sedimentation." In W. J. Weber, Jr. (ed.), <u>Physicochemical Processes for Water Quality Control</u>, Wiley-Interscience, New York, 1972.

The ordinary gravity thickener resembles a settling tank in many respects. The sludge enters the tank through the central core, is distributed radially, and exits through a sludge trough at the bottom. The clear water is discharged over a peripheral weir. The scraper mechanism assists in moving the bottom sludge toward the trough and prevents the formation of "ratholing," the formation of water cone during bottom sludge pumping.

Some thickeners have vertical picket fence posts attached to the rotating frame. These rakes are believed to assist in the consolidation process. Actually, it has been proven conclusively that stirring (at even slow speeds) is detrimental to thickening [56,57]. Only in cases where the sludge is biologically active and forms gas are the pickets of value because they promote the removal of the gas. In chemical sludge thickening, picket fence rakes are detrimental to thickening and should not be employed.

A variation of the ordinary gravity thickener is the lamella or tube settler, in which a large surface area is created by placing plates or tubes in the tank at an angle. Although this idea is not new, it significantly promotes the thickening of some slurries between the slanted plates so that it may be removed from the main body of surrounding liquid. The thickened slurry drops only a short distance and simply slides down the plate or tube. The trapped water can travel to the top on the underside of the plates. Although maintenance problems have plagued some of these systems, they have found acceptance in the thickening of such sludges as alum sludge from water treatment plants.

Sludge thickening also may take place in a lagoon. Sludge is allowed to consolidate in the lagoon, the liquid is periodically decanted, and more sludge is added as space allows. Lagoons fill up eventually and must be covered or dredged. Although lagoons appear to be an expedient means of sludge management for the short term, they pose serious long-term problems unless substantial measures are taken to prevent leakage into the subsoil.

Little information is available on the impact of metal hydroxide storage on ground water. Metal hydroxides, however, are known to clog up the pores in soil rapidly. As long as the pH remains above 7.0, metal hydroxides probably will not be a problem. In one instance, 10 yr of storage produced only a few inches of metal accumulation below the lagoon bottom [58].

The probability of accidents with stored hazardous materials should not be discounted, along with the possible litigation and adverse public relations that would result. Lagoons should be used only for short term "crisis" situations and should not be thought of as long-term solutions.

Sizing

In all gravitational thickening applications, the processes are sized according to how well the sludge settles. As the settleability improves, the required areas and volumes decrease.

Sludge settleability historically has been evaluated by tests in a 1-liter (0.26-gal) cylinder. The Sludge Volume Index (SVI), for example, is defined as the volume occupied by 1 g (0.04 oz) of sludge after 30 min of settling in a 1-liter (0.26-gal) graduated cylinder. More recently, such tests have been used to develop design criteria for thickeners.

The prevailing sludge thickening design procedure permits the calculation of the surface area required to transmit a sufficient quantity of solids through the tank per unit time. This action is referred to as solids flux and is expressed in terms of the mass of solids moved through a unit area in a given time [e.g., $g/h/m^2$ ($lb/h/ft^2$)].

The required area for thickening can be estimated by calculating the critical solids flux--the flux that will limit the operation of the thickener. A series of batch thickening tests are conducted using sludge at different solids concentrations. The height of the sludge/liquid interface is recorded with time, and the results are plotted, as in Figure 48a. The batch flux (Figure 48b) is defined as the settling velocity times the solids concentration.

$$G_B = (v)(C) \qquad (44)$$

where G_B = batch flux ($g/h/m^2$)
 v = velocity, or change in height divided by change in time, $\Delta H/\Delta T$ (m/h)
 C = concentration (g/m^3)

In a continuous thickener, the total flux consists of not only the solids settling through a given area but also the movement of the solids downward because of the removal of the underflow. Consider a thickener that has solids that do not settle at all (i.e., same density as the liquid). Obviously, this slurry will have no batch flux ($v = 0$, hence $G = 0$ for all C). Placed in a continuous thickener, however, a downward flux of solids will exist as long as there is extraction of the underflow. This flux is known as the underflow flux and is defined as:

$$G_u = (Q_u/A)C \qquad (45)$$

where G_u = underflow flux ($g/h/m^2$)
 Q_u = flow rate of the underflow (m^3/h)
 A = thickener area (m^2)

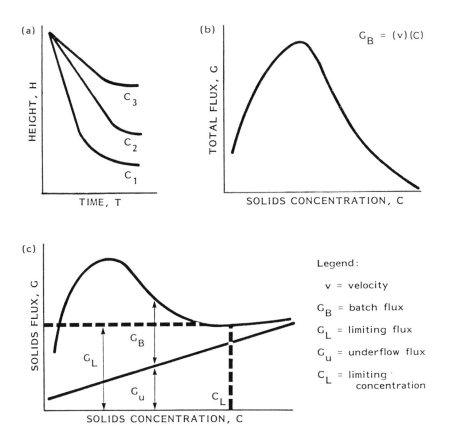

Figure 48. Calculation of limiting flux: (a) multiply velocity of sludge/liquid interface for different concentrations by corresponding solids concentration, (b) plot flux versus concentration, and (c) superimpose batch flux and determine low point in composite curve.

This underflow flux can be plotted on a graph of flux versus concentration and will be a straight line for a given Q_u and A.

Thus, the total flux, G, is

$$G = G_B + G_u \qquad (46)$$

and this can be plotted as shown in Figure 48c. If the solids concentration of the influent is C_o and the desired underflow concentration is C_u, the total flux curve passes through a minimum. This minimum is the critical flux, which determines the minimal area for the thickener because

$$G_L = Q_o C_o / A_{min} \qquad (47)$$

where G_L = limiting flux
Q_o = flow rate of influent
A_{min} = minimal area

If the loading is too high, the thickener cannot handle the solids, and some of them will be discharged over the effluent weir.

Tests with cylinders can also provide an approximation of the possible benefit of gravity thickening, that is, the concentration of the underflow, Cu. This is obtained by allowing the sludge to settle to its final volume and taking the ratio of the heights, so that

$$C_u = C_o (H_o / H_f) \qquad (48)$$

where C_o = initial solids concentration (g/m^3)
H_o = initial height of the sludge in the cylinder (m)
H_f = final compacted height (m)

A complete description of the thickener theory may be found elsewhere [56]. The important aspect of this analysis is that thickeners can only be loaded to a limit, at which point they will not accept additional solids.

It is necessary to caution against the use of small [e.g., 1-l (0.26-gal)] cylinders for the design of large, expensive thickeners. For reasons not yet adequately understood, small cylinders exert a scale effect on sludge settling. As shown in Figure 49, cylinders shorter than 0.9 m (3 ft) tend to produce lower settling velocities [59], and cylinders less than 0.3 m (1 ft) in diameter tend to increase or decrease the velocity, depending on the sludge and the concentration [60]. Additional problems with batch testing include methods of filling the cylinder and other laboratory artifacts.

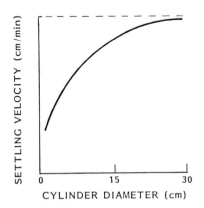
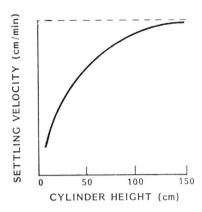

Figure 49. Effect of initial depth and cylinder diameter on settling velocity for a typical activated sludge at 2,000 mg/l.

The testing of chemical sludges is even more complicated because the quality of the sludge can change markedly with time and method of analysis. For example, a metal hydroxide sludge such as zinc hydroxide [$Zn(OH)_2$] can be produced in the laboratory by raising the pH of a zinc solution to about 10. The precipitate forms readily and will settle in a cylinder. Unfortunately, the settling velocity is dependent on such variables as how rapidly the pH was increased, how much stirring was involved, whether the pH was raised and then lowered, and the age of the sludge. The last variable, sludge age, is an especially difficult problem because the speciation of oxides and hydroxides is changing in the first few minutes of the settling test, so the sludge at the end of 10 min is very different from what it was in the beginning.

It is also evident that metal oxide/hydroxide sludges settle at different rates, as shown in Figure 50. When the sludges are mixed, the resulting sludge tends to settle at the rate of the slowest of the two. This fact has been recognized for years in municipal wastewater treatment, where the addition of a well settling primary sludge to a poorly settling waste-activated sludge produced a poorly settling mixed sludge. The mixture of sludges presents a major problem in design because most waste sludges from metal finishing are mixtures of many metals. In one survey, only 11 percent of the waste sludges included one metal, whereas 17, 20, and 53 percent had two, three, and four or more metals, respectively [60].

Economics

A gravity thickener is a fairly inexpensive piece of equipment, as seen by the estimated construction cost curve shown as Figure 51. The operation and maintenance cost is similarly low because the only energy required is for the rake arm mechanism motor and the possible pumping of the slurry. The construction costs in this figure (as in all subsequent figures showing capital costs) are based on an EPA-sponsored study conducted in 1976 and include all the necessary ancillary equipment, materials, and installation costs for constructing a thickener [61]. The construction curve is thus an estimate of the total capital outlay for a thickener. The operation and maintenance curve, expressed as an annual cost, includes power [estimated at $13.89/GJ ($0.05/kWh)], labor ($10/h), and other operation and maintenance costs.

Because of the highly variable land costs and the nature of sludge behavior in lagoons, it is not reasonable to estimate the costs of sludge disposal in lagoons. As a frame of reference, however, in 1976 dollars a lagoon costs about $0.81/m^3 ($1,000/acre-ft) of lagoon volume [61].

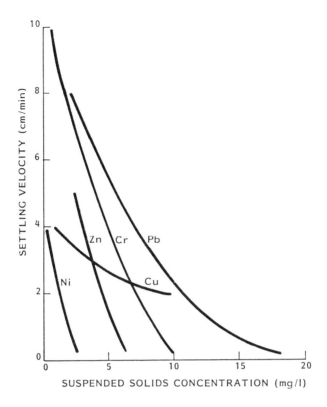

Figure 50. Representative settling data for metal hydroxide sludges.

FLOTATION

Wastewater sludges tend to be very light and thus settle and compact poorly under the force of gravity. For example, waste-activated sludge flocs have densities of less than 1.08 (specific gravity of water equals 1.0). Accordingly, it is sometimes easier to achieve separation of solid flocs from liquid by promoting their flotation.

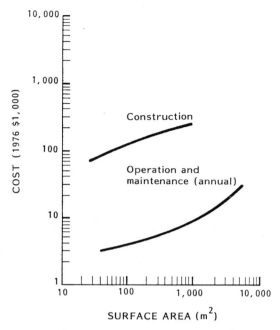

Note.--Labor is based on $10/h and power is based on $0.05/kWh.

SOURCE: G. L. Culp, Handbook of Sludge Handling Processes, Garland STPM Press, New York, 1979.

Figure 51. Estimated capital cost of a gravity thickener.

In a flotation process, suspended particles in the wastewater are made to float to the surface of the water where they can be skimmed off. Small gas bubbles are introduced into the wastewater. The gas bubbles may either become entrapped in the particle structure or adhere to the particle surface. The buoyant force of the combined particle and gas bubble causes the particle to rise to the surface. Flotation has two advantages over sedimentation. First, small or light particles that settle slowly can be removed completely and in a shorter time. Second, flotation can achieve equivalent efficiencies at higher surface loadings and, therefore, requires less space. Four methods are available for introducing gas bubbles into wastewater: air flotation, dissolved air flotation, vacuum flotation, and electroflotation [62].

The air flotation method involves the introduction of air into the liquid through a revolving impeller, which breaks up the air stream into small bubbles and disperses the bubbles throughout the water. This method is used more often to concentrate minerals from ores than to treat wastewater [36].

In dissolved air flotation, air is dissolved in the water under pressure; when the pressure is released, the air is released from solution as bubbles that have average diameters of 60 to 80 µm (2.4 to 3.2 X 10^{-3} in) [63]. The action is comparable to opening a bottle of a carbonated beverage. Various degrees of pressure may be used depending on the nature of the waste. Full-flow pressurization involves pressurizing the entire wastewater stream, whereas partial pressurization affects only a portion of the wastewater stream and mixes it with the unpressurized portion of the wastewater as it enters the flotation tank. With recycle pressurization, a portion of the clean effluent from the flotation tank is pressurized and the clean, air-saturated stream is mixed with the wastewater as it enters the flotation tank. Recycle pressurization usually is applied to wastewater that is chemically treated to aid in floc formation because the shearing stresses in the pumps and pressurizing system would destroy the floc. Because wastewater containing heavy metal suspensions usually is treated chemically to promote flocculation, dissolved air flotation with recycle pressurization is the flotation process that should be used for clarification. Figure 52 illustrates a dissolved air flotation system with recycle pressurization.

Vacuum flotation saturates the wastewater with a gas at atmospheric pressure and then applies a vacuum. Under the vacuum, the solubility of the gas in the liquid is decreased and the gas is released from the solution as small bubbles. Vacuum flotation is not used as often as dissolved air flotation because it produces a lower quality effluent and has maintenance difficulties [48].

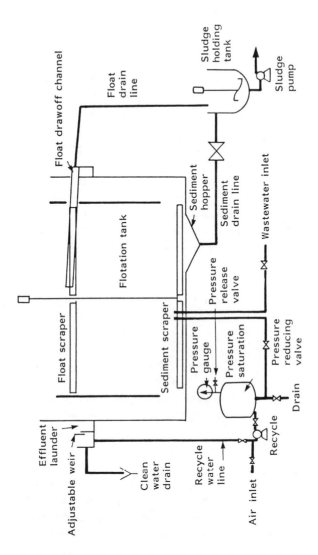

Figure 52. Dissolved air flotation system with recycle pressurization.

Electroflotation involves the generation of bubbles by electrolysis. An electroflotation unit consists of a rectangular tank with a pair of electrode grids near the bottom. The cathode is placed above the anode; when direct current flows between them, oxygen bubbles form at the anode and hydrogen bubbles form at the cathode. As the bubbles rise to the surface, they carry suspended particles. Electroflotation has been found to be a practical treatment method for removing suspended solids, fats, and biochemical oxygen demand (BOD) from industrial wastes in the United Kingdom [62].

Design

The factors influencing the efficiency of flotation for suspended solids removal include the degree of adhesion between the particles and the air bubble present and the buoyancy of the particle with the air bubble attached. These factors determine the amount of gas needed to float the solids as well as the stability of the floated solids [62]. Because these factors vary with the physical and chemical characteristics of the suspended particles in the wastewater, laboratory and pilot-plant testing usually are required to establish the design basis for a flotation unit.

The flotation characteristics of a waste can be estimated by use of a laboratory cell (Figure 53). The calibrated cylinder is filled with the wastewater mixture and the pressure chamber is filled with clarified wastewater.

Compressed air is applied to the pressure chamber to a pressure of 380 to 450 kPa (40 to 60 $lb/in^2 g$). The air/liquid mixture is shaken for 1 min and allowed to stand for about 3 min to attain saturation at the desired pressure. A volume of the pressurized effluent is released into the calibrated cylinder to mix with the wastewater. The volume released is related to the recycle ratio of a pressurized recycle flotation system. The release rate should be controlled to avoid shearing of the suspended solids in the wastewater and to maintain adequate mixing. The rise of the sludge interface is measured over time. After a 20-min detention time, the clarified effluent and the floated sludge are drawn off and their suspended solids concentrations are measured.

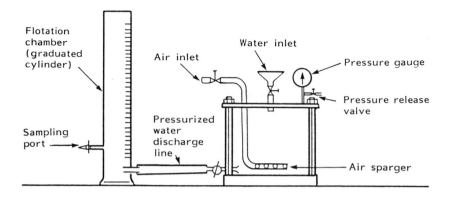

Figure 53. Laboratory flotation apparatus.

The experiment is repeated for different ratios of air to wastewater solids to determine an optimal value. The air-to-solids ratio for pressurized recycle is calculated from the following equation [64]:

$$\frac{A}{S} = \frac{1.3[s_a R(fP - 1)]}{QS_a} \quad (49)$$

where s_a = air saturation, at atmospheric pressure (cm^3/l)
  R = pressurized volume (l)
  P = absolute pressure (atm)
  Q = waste volume (l)
  S_a = influent suspended solids (mg/l)
  f = efficiency of saturation
  1.3 = a conversion factor that makes the ratio unitless (mg/cm^3-atm)

For the test described above, $f = 1$. In an industrial flotation system, f may vary between 0.5 and 0.8 [64,65].

The surface area required for the flotation unit will depend on the time required for the suspended solids to rise to the surface and the degree of float compaction desired. In practice, flotation cell areas of 0.37 to 1.4 $m^2/l/s$ (15.08 to 57.04 $ft^2/gal/s$) are used [48]. Process efficiency decreases with decreasing area per unit of flow rate. The design flow rate is the sum of the wastewater flow rate and the recycle flow rate.

Equipment

The principal components of a dissolved air flotation system include the system for saturating the recycle stream with air under pressure and the flotation unit.

A number of methods for saturating the recycle stream with air have been reported. Air may be added to the recycle stream through an injector on the suction side of the recycle pump. The air and water are mixed as they pass through the pump and are retained under pressure in a tank with a detention time of 1 to 3 min. A back-pressure regulating device maintains a constant head on the pump. With this method, between 35 and 45 percent of the injected air will dissolve in the water, and up to 50 percent saturation can be achieved [66]. The recycle stream may also be saturated with air by passing the water through a packed column countercurrently to the flow of air through the column. This method can produce 90 percent saturation.

A novel method for dissolving the air into the wastewater under pressure, called microflotation, was developed in Sweden. In this process, the wastewater flows down an 8- to 15-m- (25- to 50-ft-) deep shaft countercurrently to the flow of air entering at the bottom of the shaft. Because of the increase in hydrostatic pressure as the water flows downward, increasing amounts of air dissolve in the water. As the water rises on the other side of the shaft into the flotation tank, hydrostatic pressure declines and the dissolved air is released as microscopic bubbles [62].

The flotation unit may be either circular or rectangular with a skimming device to remove the floated sludge and a bottom scraper to remove any settled sludge. The flotation tank must permit the air and suspended solids to rise to the surface of the tank with a minimum of interference (e.g., turbulence or obstructions).

Effluent ports must be sufficiently submerged to prevent interference with the froth on the surface. Baffles, walls, or other obstructive devices at the inlet to the flotation tank tend to destroy aggregate bonding and reduce flotation efficiency. Turbulence in the region of the froth will result in losses of floated solids [49]. Outdoor flotation tanks may require wind screens or high freeboards to minimize turbulence caused by the wind blowing across the surface of the tank.

Operation

Flotation thickeners have been used for a number of years in the thickening of waste-activated sludge. The flotation unit requires that air be dissolved under pressure and then mixed with the incoming slurry. As the pressure is released, the air bubbles attach themselves to the sludge particles, increase their buoyancy, and cause the particles to travel to the top where the thickened solids are skimmed off.

In almost all cases, the sludges thickened by flotation are conditioned with chemicals such as organic polyelectrolytes. Such "polymers" tend to agglomerate the particles and create discrete flocs, which can then be buoyed upward. Tiny particles, although light, require too much time to rise and would not be removed.

Sizing

The amenability of a sludge to be thickened by flotation can be tested by a simple setup illustrated in Figure 53. As the compressed air in water is released into the sludge, the sludge is mixed, and the small bubbles are attached to the solids, which begin to rise. The rate of the sludge/liquid interface is recorded, and its velocity is exactly analogous to the settling velocity in a batch thickening test. The required thickening area can be calculated in a like manner.

The extra variable present is, of course, the amount of air and its pressure. The effect of the air is best described by the ratio of the air to the solids (weight to weight). The effect of the air/solids ratio is described in Figure 54. Note that as the amount of air is increased for a given solids, the recovery as well as the thickened solids concentration achieve a practical limit, and there is nothing to be gained by adding more air to the thickener. The most efficient operating point can be ascertained by conducting a series of laboratory tests at various air/solid ratios.

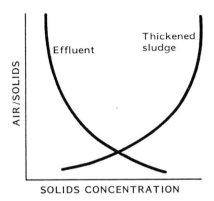

Figure 54. Dissolved air flotation system: air/solids ratio affects both thickened solids and solids recovery.

The apparatus pictured in Figure 53 is subject to the same potential laboratory artifacts as discussed under gravitational thickening. Further, the laboratory system does not accurately reflect the actual operation of a dissolved air flotation thickener because the air is mixed with the sludge as a batch operation all at once, whereas in a full-scale thickener the water containing the dissolved air is continuously mixed with the sludge and the pressure is released on the mixture. This operation can be simulated by blending the sludge and water into the flotation cylinder at the required ratios; more accurate results can be obtained with such a modification [67].

Economics

In conducting flotation tests with metal hydroxide sludge, the same problems with sludge stability apply as noted earlier. Although no data have been published on flotation of these sludges, it seems reasonable that an additional problem would be the large availability of oxygen and the changes in sludge characteristics that may cause changes in speciation.

The economics of a dissolved air flotation thickener are illustrated in Figure 55. It should be reemphasized that, in most wastewater treatment plants, chemical conditioning of sludge has been found to be necessary for flotation to be successful. This may not apply to the flotation of metal hydroxide or other chemical sludges because few data are available. As before, the construction cost is an estimate of total capital outlay for the flotation system, whereas the operating and maintenance costs include power and labor.

Because of rapidly increasing electrode and rectification cost with volume of waste to be treated, electroflotation usually is limited to applications where flows are less than about 20 m^3/h (700 ft^3/h). The cost for an electroflotation unit handling this size stream was reported to be between $20,000 and $30,000 in 1976, about the same as a microflotation plant of the same capacity [62]. Power requirements for electroflotation were reported to be 0.4 kWh/m^3 (38.6 Btu/ft^3) [62].

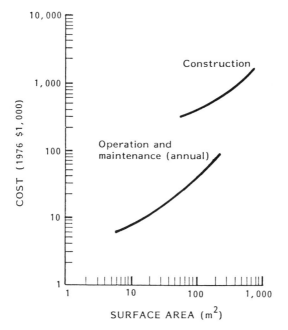

Note.—Labor is based on $10/h and power is based on $0.05/kWh.

SOURCE: G. L. Culp, Handbook of Sludge Handling Processes, Garland STPM Press, New York, 1979.

Figure 55. Economics of dissolved air flotation.

7
Sludge Dewatering and Disposal

This section covers the two main operations associated with the management of chemical sludges: concentration of the solids to remove extraneous water and the ultimate disposal of chemical sludges into the environment.

Solid/liquid separation, as discussed in Section 6, dealt with thickening--the process of removing water to the point where the remaining sludge solids still behave as a liquid. This section considers centrifugation, filtration, and drying--the solid/liquid separation processes that remove water to the point where the product behaves as a solid.

At times, sludges can be disposed of into the environment without further treatment. For example, the spraying of waste-activated sludge or sludges from canning plants into forests is a well-established practice. Such sludges are readily assimilated by the forest ecosystem and cause no further adverse environmental impact. Sludges can also be used in other chemical processes or as an additive in industrial products.

Too often, however, it is not economically or environmentally feasible to dispose of such sludges directly into the environment or to use them for another beneficial purpose. Raw primary sludge, for example, is odoriferous, contains large solids and pathogenic organisms, and can seldom be discharged onto land without some prior treatment. Similarly, sludges containing high concentrations of metal hydroxides can be toxic to plants and subsequent consumers and cannot be disposed of into a natural ecosystem without causing serious injury. Accordingly, many waste sludges must be transported to secure landfills or treated further to facilitate ultimate disposal. Such treatment may involve the removal of some of the liquid (to make transportation less expensive), the detoxification of sludge, or even incineration.

Incineration is seldom used because the majority of solids in metal finishing sludges will not burn. High temperature roasting has been used, however, to convert the metal hydroxides to oxides. This treatment reduces the volume and solubility of the sludge, which makes it more amenable to disposal [66]. Precautions must be taken, however, to prevent air emissions of volatilized metals.

The types of detoxification can be accomplished by mixing the sludge with other chemicals so that:

- The product is nontoxic (for example, neutralization).
- The toxic materials are so well fixed within the sludge that they cannot be discharged into the environment and cause subsequent damage (for example, chemical fixation).

SLUDGE DEWATERING TECHNIQUES

Centrifugation

As with gravitational and flotation thickening, the process of centrifugation is the separation of solids from liquid according to density. The difference is that thickening uses gravitational force, whereas centrifuging multiplies gravity by as much as 1,000 times. Nevertheless, it should be kept in mind that centrifugation is no different in principle from thickening, and most of the problems of scale-up and operation apply to both processes.

Operation

The three most widely used centrifuges for wastewater sludge treatment are the basket, disc, and solid bowl or decanter.

The basket centrifuge, shown in Figure 56a, operates on a cyclic instead of continuous principle. The sludge is pumped into the spinning bowl and, as the solids move by centrifugal force, to the inner wall. The centrate (clear

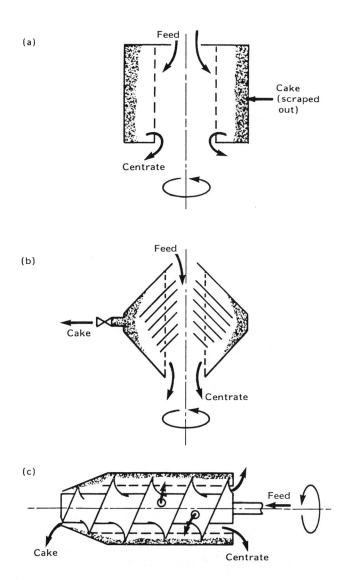

Figure 56. Centrifuges used for wastewater sludge treatment (a) basket, (b) disk, and (c) solid bowl.

liquid) is discharged out the overflow weir. Eventually, the basket will fill up with solids, which must be scraped out. In some models, an intermediate step involves the skimming off of the light, poorly compacted solids that are wasted with the centrate. This is usually not a good idea in wastewater treatment because these solids must be removed eventually and will cause future dewatering problems or they will be discharged with the plant effluent. A basic principle of sludge management is to achieve the greatest solids recovery possible, even at the expense of sludge solids concentration.

Modern basket centrifuges are fully automated and can run without operator supervision. The problem of cyclic feed from a continuous flow is often solved by installing two or more machines in parallel so that while one is being emptied, the other is being fed.

The second type of machine used in wastewater treatment is the disc centrifuge. Originally invented for the separation of cream from milk, the disc centrifuge uses the same basic principle as the lamella or tube settlers. In the disc machine, a series of concentric cones (discs) are stacked inside so that the settling occurs within these layers and the sludge slides to the outside. As shown in Figure 56b, the sludge solids are collected in the far inside wall. The removal of these solids presents a design problem. In some models, the entire centrifuge casing opens up for a short time and the solids are discharged. In other models, a valve is installed that opens periodically. A modern version of the latter machine has a recirculation loop that recycles the sludge to the inside of the machine, and the excess sludge is bled off as required.

A major drawback in the application of disc machines to municipal wastewater treatment has been clogging. Municipal sludges can contain all manner of solids, and the small spaces between the discs are susceptible to clogging. Once clogged, the machine must be disassembled--a long and tedious task.

The third type of centrifuge, widely used in the dewatering of chemical and wastewater sludges, is the solid bowl or decanter centrifuge. Illustrated in Figure 56c, this machine consists of a solid, tapered bowl that is rotated on its longitudinal axis. Typically, the large end of the bowl has holes that allow the centrate to escape, whereas the tapered end, or beach, has holes for sludge solids removal. As the solids settle to the inside wall, a screw conveyor, which rotates slightly slower than the bowl, moves the solids to the open end and out of the bowl. The feed is introduced by means of a central feed pipe (not rotating) that sprays the sludge into the machine. For some time it has been recognized that the introduction of the feed is a critical component of successful centrifugation. The feed tends to splash into the rapidly rotating sludge and destroy some of the separation that has already occurred. Further, it is necessary to accelerate the feed from 0 to perhaps 4,000 r/min in a short time. The typical hydraulic residence time in a solid bowl centrifuge is about 20 s. The recognition of this problem has prompted the development of newer machines, which run at slower speeds and introduce the sludge to the bowl by accelerating it first with an inner cone. These machines have found wide acceptance in the dewatering of light and fluffy sludges, such as activated sludge, alum sludge, and other chemical sludges.

The design and operation of a solid bowl centrifuge must consider a number of machine and process variables. Machine variables include the following:

- pool depth
- bowl length
- beach angle
- bowl speed
- conveyor speed
- number of leads and pitch of conveyor
- bowl radius
- point of chemical addition.

The operator may be able to change some of the machine variables, such as pool depth and bowl speed, but this is seldom done. The most important variables at his command are the process variables, which include flow rate and such feed characteristics as solids concentration, chemical conditioning, and age of sludge.

The centrifuge has two objectives: to produce a dry sludge cake and to discharge a clear centrate. Unfortunately, the two objectives are difficult to attain simultaneously. In fact, minor adjustments in any of the variables (with the exception of sludge characteristics) produce data which, when plotted on a graph of solids recovery versus cake solids concentration, will produce one line (Figure 57). For example, increasing the pool depth (thus decreasing the dry beach time and increasing the hydraulic residence time) increases the solids capture. Because the residence time is greater, a larger fraction of the solids can settle out. The increase in solids capture, however, is achieved at the expense of cake solids. The solids have less time to dewater on the beach, so the wetter solids particles are captured and the water they carry contributes to the wetter cake.

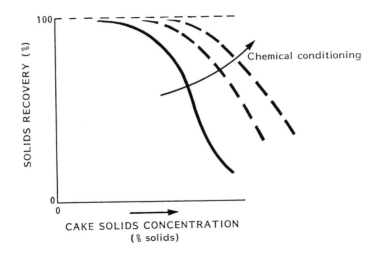

Figure 57. Centrifuge performance trade-off between cake solids concentration and solids recovery.

Only by changing the characteristics of the sludge can both recovery and cake solids be increased (shift the line in Figure 57). The most common means of altering sludge characteristics is to condition the sludge chemically, usually with organic polyelectrolytes. Another way of changing the sludge characteristics is to adjust the sludge processes that precede the centrifuge. For example, a raw primary sludge in wastewater treatment dewaters fairly well with centrifugation but will not dewater as well if it has become septic. The age and condition of the sludge can have a significant effect on performance. The problems with sludge production and the time-variable properties of metal oxides (discussed under thickening) also apply to chemical sludges.

Sizing

Centrifuges are basically highly efficient thickening tanks; therefore, it is not surprising that centrifuge sizing is based on surface area and hydraulic retention time. The most widely used technique for sizing centrifuges is the Sigma concept, which was first developed in 1952 as a means of scaling up geometrically similar centrifuges [68]. Assuming that the particles in the centrifuge settle in the laminar flow regime, exhibit unhindered settling, and accelerate immediately to rotational speed upon introduction into the centrifuge (all untenable assumptions), it can be shown that the flow rates at equal performance[a] for two geometrically similar solid bowl machines can be related as:

$$Q_1/Q_2 = \Sigma_1/\Sigma_2 \qquad (50)$$

where $\Sigma = V\omega^2/[g(\ln)(r_2/r_1)]$
 V = hydraulic volume in the bowl
 ω = rotational speed (rad/s)
 g = gravitational constant
 r_1, r_2 = radii from bowl centerline to the surface of the sludge and inside bowl wall, respectively

---

[a] Some authors call this the centrifuge capacity. In truth, a centrifuge does not have a liquid-handling capacity. At higher flow rates, performance simply suffers.

Sigma has different formulations for other types of centrifuges.

For a solid bowl machine, as with other centrifuges, another limitation on performance is the solids loading. If a machine is loaded with solids at a rate higher than the conveyor can move them out, there is a solids buildup in the bowl, reducing the hydraulic volume and adversely affecting performance. The solids loadings can also limit a centrifuge, and the relationship between any two geometrically similar machines can be related using the Beta concept, where

$$Q_1/Q_2 = \beta_1/\beta_2 \tag{51}$$

where $\beta = \pi(\Delta\omega)SNDz$
- $\Delta\omega$ = speed difference between bowl and conveyor
- S = pitch of conveyor
- N = number of leads
- D = diameter of bowl
- z = pool depth

The performance of a centrifuge can be limited by solids loading as well as by liquid rate. Figure 58 shows solids recovery versus liquid feed rate. These data were obtained using a calcium carbonate slurry and show that, at a given flow rate, solids recovery cannot be maintained at a higher percentage of solids. At a 20-percent solids concentration, recovery becomes solids limited.

When data from a smaller, geometrically similar machine are not available, it is necessary to estimate the required size by laboratory tests. The most common test is to centrifuge the sludge in a test tube desk-top machine, judge the clarity of the centrate, and estimate the cake consistency by poking the sludge cake with a glass rod. Both characteristics are important because the particles must first settle out and then be conveyed by the screw. A well settling sludge that produces a very soft cake is not a good candidate for solid bowl centrifugation.

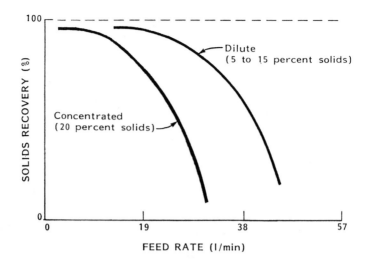

Figure 58. Solids recovery of a dilute and concentrated calcium carbonate slurry.

These tests can be quantified to allow for the comparison of different sludges and to detect changes in a sludge with time.

The settleability of a sludge in a test tube can be measured by using a strobe light that is synchronized with the spinning test tubes (Figure 59). If the shield holding the test tube has a hole in it, it is possible to observe the sludge/liquid interface and determine its height with time. This information yields the velocity of sludge settling at a given centrifugal acceleration and a given solids concentration. The analysis described earlier for estimating the required surface area of a thickener can be applied here for the calculation of centrifuge surface area [56].

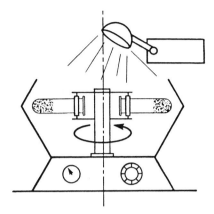

Figure 59. Use of strobe light to observe settling of sludge under gravitational force.

It is possible to use this test to develop a measure of sludge settling characteristics under a centrifugal acceleration. The settleability coefficient has been defined as:

$$S = v/\omega^2 r \tag{52}$$

where v is the velocity of the interface at some distance, r, from the centerline, rotating with a centrifugal speed of ω rad/s. The value of S seems to be independent of the centrifugal force imposed; therefore, this is a true measurement of how well the sludge settles [69].

The characteristic of sludge firmness or body can be evaluated with a penetrometer, shown in Figure 60. A metal or plastic rod is dropped into the sludge cake and its penetration is measured. A sludge that will move readily in a scroll centrifuge cannot be easily penetrated, whereas a light sludge (such as metal hydroxide slurries) will be soft, penetrable, and difficult to move out of a solid bowl centrifuge [70].

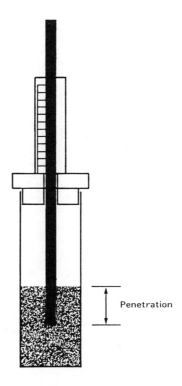

Figure 60. Use of penetrometer to estimate firmness of sludge.

Economics

It is difficult to estimate the economics of sludge centrifugation because the performance must be defined. Figure 61 is an estimate of the capital and operating costs of centrifuges used in wastewater treatment, but a wide variation is expected [61].

Sludge Dewatering and Disposal   171

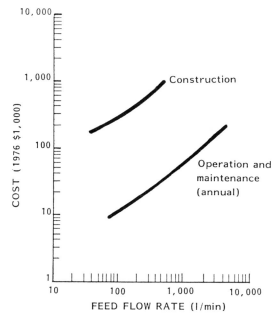

SOURCE: G. L. Culp, Handbook of Sludge Handling Processes, Garland STPM Press, New York, 1979.

Figure 61. Costs of sludge centrifugation.

Filtration

The filtration of sludges involves the use of a porous medium, such as a fabric, that allows the passage of liquids but not solids. In some cases, such as with a gravity filter, no additional force is employed. Most filters, however, use either vacuum or positive pressure.

Operation

The most common vacuum filter is the rotary drum filter, shown in Figure 62a. Modern rotary vacuum filters pick up sludge from a bottom trough, dewater it through a fabric (or metal coil) covering the perforated vacuum

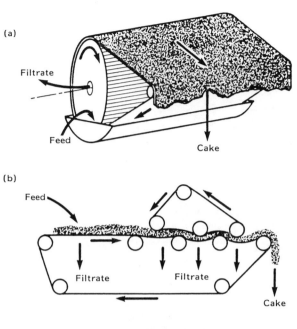
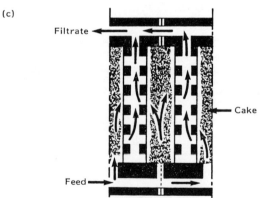

Figure 62. Types of sludge filters: (a) rotary drum vacuum filter, (b) belt filter, and (c) filter press.

drum, and drop it off as the fabric is lifted off the drum and forced to go over a small roller.

The standard vacuum filter does not perform well with light sludges, such as activated sludge and some metal hydroxides, even after chemical conditioning. The sludge will not be picked up by the filter cloth, and the pores of the fabric are plugged too rapidly. Plugging of the pores is known as blinding. A precoating technique can improve the performance but represents a substantial operating expense.

The belt filter, first developed in Europe and introduced in the United States only 10 yr ago, is a marked improvement over the rotary vacuum filter for dewatering difficult sludges. Although the various proprietary systems differ in detail, most belt filters employ the basic mechanisms illustrated in Figure 62b. A chemically conditioned sludge first drops on to a perforated belt, where gravity drainage occurs. The already thickened sludge is pressed between a series of rollers to produce a dry cake. A major problem with some sludges is the tendency to squirt out the sides of the belt as it is being squeezed.

In the filter press (Figure 62c), a chemically conditioned sludge is pumped into cavities that are formed by a series of plates covered by a filter cloth. The liquid finds its way through the filter cloth, leaving the sludge solids behind. The solids fill up the cavity eventually and become the sludge cake. The plates are opened up and the sludge is removed.

This cyclical process is not popular in wastewater treatment because it requires operator attention, especially during cake discharge. It does, however, exhibit one major attribute: given a sufficiently long filter run, it will dewater most sludges and will produce a cake dryer than that attainable by any other mechanical process.

Sizing

Filters are sized according to "filter yield," which is defined as the dry solids produced as cake per unit time per unit area of filter surface.

Typically, filters used for dewatering municipal sludge produce dry solids at a rate of 32 to 64 kg/h/m^2 (2 to 4 lb/h/ft^2).

Filtration on a rotary vacuum filter involves sludge pickup, sludge dewatering by filtration, and sludge cake discharge. These steps can be readily duplicated using a filter leaf apparatus, illustrated in Figure 63.

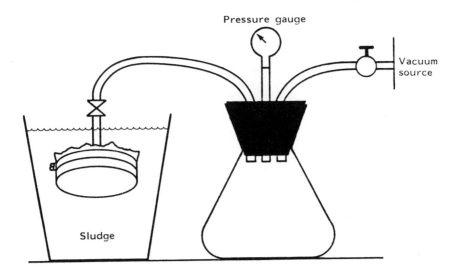

Figure 63. Filter leaf apparatus used for sizing vacuum filters.

A grooved disc, covered with a filter fabric and connected to a vacuum source is immersed into sludge for a specific time and then removed (sludge pickup) and held up in the air (dewatering). The vacuum is released and the cake scraped off (sludge cake discharge). The cake solids are dried and weighed, and the amount of dry sludge is determined. The filter yield is calculated using the weight of solids and the area of the filter leaf and the time of the total filtration cycle.

At present, no similar sizing techniques have become standard practice for belt or vacuum filtration although some work has been done on both. For the belt filter, a simulation technique involving the pressing between two layers of fabric has been developed in England [71]. Pressure cells simulating pressure filters have been used by some researchers [72].

In all filtration applications, it is advantageous to be able to characterize a sludge by its ability to be filtered, which is analogous to the settleability coefficient used for characterizing the ability of a sludge to dewater by centrifugation. The concept of specific resistance to filtration has gained wide acceptance as a means describing the filterability of sludge. As the specific resistance increases, the sludge becomes more difficult to dewater.

The apparatus shown in Figure 64 is used for measuring specific resistance to filtration. Sludge is poured into the Buchner funnel lined with a paper filter, and the rate of filtrate production is measured in the graduated cylinder. The filtrate volume is plotted against the time divided by the volume (Figure 65). This usually results in a straight line with a slope, b. The specific resistance, r, is calculated as

$$r = 2PA^2 b/\mu w \tag{53}$$

where P = vacuum pressure employed
  A = area of Buchner funnel
  b = slope
  µ = viscosity
  w = cake deposited per volume of filtrate

The measurement of specific resistance is somewhat cumbersome, especially if a series of tests is needed for estimating the effect of chemical conditioners. Such a series could run into hundreds of tests. A "quick-and-dirty" technique has been developed for estimating sludge filterability, based on the movement of water out of a sludge and into a blotter paper. As illustrated in Figure 66, this device measures the time necessary for the water in the blotter to move 1 cm (2.5 in) and the time can range from a few seconds for extremely well conditioned sludge to several minutes for sludges that do not release their water readily. The capillary suction time (CST) is a useful tool for rapidly estimating the ability of a sludge to release its water. It is not, however, a measure of how well a specific sludge would filter with a given filtration system because other factors, such as sludge pickup, clogging of the filter cloth, and discharge, play equally important roles.

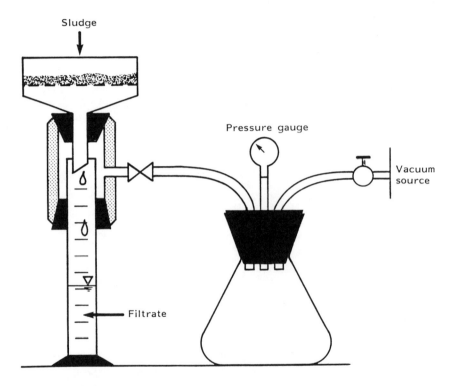

Figure 64. Buchner funnel apparatus for determining specific resistance to filtration.

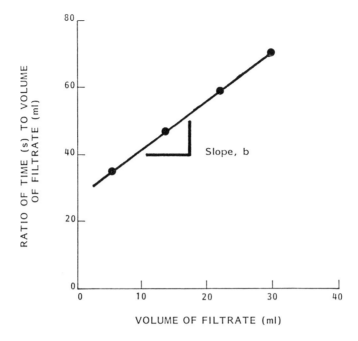

Figure 65. Typical results of a Buchner funnel test for measuring specific resistance to filtration.

Economics

The costs of three types of filters are shown in Figure 67. In all cases, the capital costs include the entire filtration system, with chemical feed pumps, tanks, and an enclosure.

Drying

Mechanical dewatering devices can only remove free and floc water. Further dewatering must be accomplished by thermal means. Most drying devices used on sludges are too expensive and some are too dangerous. A sludge high in organic materials (e.g., activated sludge) when dried forms an extremely combustible powder. The number of explosions in drying plants has had a serious effect on the use of thermal drying in wastewater treatment. With inorganic sludges, such as metal hydroxides, the explosion potential is negligible, but the cost of energy remains a detriment to the use of thermal drying.

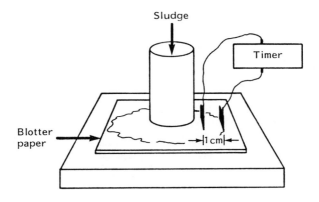

Figure 66. Capillary suction time apparatus.

Operation of Sand Drying Beds

Figure 68 illustrates a typical sand bed. The sand covers a tile field and the sludge is spread on the sand, usually about 20 cm (8 in) deep. During the first few days, the primary means of dewatering is filtration through the sand. As the pores are clogged up, however, the major liquid removal method becomes evaporation. Typically, sludges lose enough liquid to crack, which promotes further drying. In 30 to 60 d, municipal sludges are usually at solids concentrations of 30 to 40 percent. Covered beds can yield even drier solids if necessary.

Each sludge dries differently on sand beds. Alum sludge, for example, tends to form a hard, dry crust on the top, thus trapping all the liquid below. Such sludges cannot be readily dewatered on sand beds.

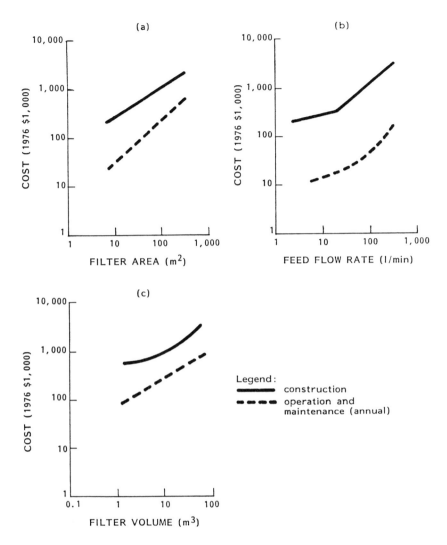

SOURCE: G. L. Culp, Handbook of Sludge Handling Processes, Garland STPM Press, New York, 1979.

Figure 67. Economics of sludge filters: (a) rotary drum vacuum filters, (b) belt filters, and (c) pressure filters.

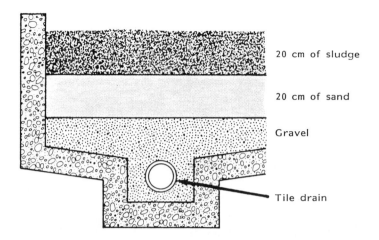

Figure 68. Cross section of a standard sand drying bed.

Sizing of Sand Drying Beds

The preferred means of sizing sand beds is to construct a model bed and test it. The model should be large enough to reduce wall effects and preferably should be sunk into the ground to minimize temperature fluctuations. A cover should be placed over the bed that will prevent rain from falling into the sludge but will not affect air circulation. This cover will eliminate the rainfall variable, which can be factored in during the final design.

Economics of Sand Drying Beds

Because of the large variability in the time necessary to dewater sludges on sand beds, it is not realistic to estimate costs. As a frame of reference, however, the capital cost of a sand bed in 1976 dollars is about $32/m^2 ($3/ft^2) of surface area [61].

SLUDGE DISPOSAL

Two sludge disposal issues that often conflict are economics and environmental impact. Each factor includes related constraints. For example, the ease of sludge disposal and uninterrupted operation are important to industries. Similarly, the long-term effects and liabilities of sludge disposal are linked closely with environmental concerns and the associated regulations.

One means of disposing of sludges is to give them to another industrial concern that would use it as new material. Waste clearinghouses have been used successfully in Europe and the United States to facilitate the exchange of industrial wastes.

In most cases, however, it is not economically or environmentally feasible to dispose of such sludges directly into the environment or to use them for another beneficial purpose. Disposal methods, especially for metal-bearing industrial sludges, are limited. Because ocean disposal is prohibited by law, and disposal into other watercourses is restricted, disposal on and under the ground is the only alternative. For industrial sludges, the available disposal options are:

- landfill
- landfill with prior treatment, such as fixation or encapsulation
- deep well injection
- contract with a disposal firm.

Landfill

Sludges containing hazardous or toxic materials must be disposed of in secure landfills. These landfills are designed so that the wastes enclosed will not seep into the ground water. Secure landfills usually include a clay liner or a rubberized liner [e.g., polyvinyl chloride (PVC), polyethylene] to prevent seepage into the ground water. Monitoring wells are often required for periodic testing of the ground water around the landfill for potential contamination. In some landfills, the leachate (liquid generated by the landfill) is collected in a tile drainage system and treated.

The capital and operating costs of landfills are site dependent. Typically, a municipal landfill costs about $1,100 to $2,200 per metric ton per day ($1,000 to $2,000 per short ton per day) to construct, whereas the cost of

operation ranges from $2.2/metric ton ($2/short ton) to well over $22/metric
ton ($20/short ton) [73]. Costs for constructing and operating secure land-
fills have not been published, but they will be substantially higher than
those for municipal solid waste.

Landfills With Encapsulation and Fixation

Secure landfills are required for sludges containing hazardous materials, even if encapsulated in containers. To ensure that containers will not leak, they should be designed for use other than storage. For most applications, encapsulation is extremely expensive and not economically feasible. Further, the placement of toxic materials in uncontrolled environments, such as municipal landfills or private disposal sites, does not guarantee that in later years the containers will not be broken by unsuspecting individuals.

Unlike encapsulation, chemical fixation appears to be an economically attractive alternative for the disposal of many sludges. As with encapsulation, chemically fixed sludges containing hazardous materials must also be disposed of in secure landfills and not municipal landfills. Although many proprietary processes are available, they all operate on the same principle (Figure 69). The sludge from one lagoon is pumped to a mixing tank, where a fixation agent is added. This additive can be a combination of materials, including sodium silicates, portland cement, lime, and other chemicals. Hundreds of patents have been issued for the various combinations of fixative agents.

The sludge from the mixing tank is pumped to another lagoon, where the slurry sets into a solid. Whatever hazardous, toxic, or difficult-to-handle materials may have been combined in the sludge become fixed in a solid mass.

With municipal sludge, there is little if any concern for the leaching of metals out of the fixed solids. All published test results indicate minimal leaching. This may not be true for some fixatives and for some sludges. As a result, it may be necessary to conduct a laboratory study to estimate the amount of material leaching out of the fixed sludge.

The cost of chemical fixation varies widely. Typical published costs for fixation of various types of chemical sludges range from about $11 to $33 per metric ton ($10 to $30 per short ton) of solids fixed [74].

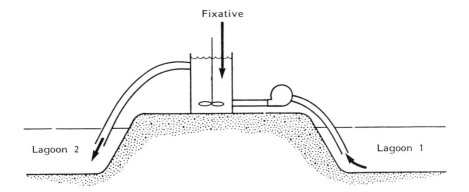

Figure 69. Chemical fixation process.

Deep Well Injection

In the past, sludge disposal has been accomplished by injecting the slurry into deep wells and storing it below ground. This technique is especially attractive in cases such as oil fields where deep wells are already available. Another advantage of deep well injection is that it is a final waste disposal step because future contact with humans is unlikely.

Deep well injection has not always been a popular disposal technique. The injection of toxic and otherwise hazardous chemicals into lower geological strata involves considerable risk and is, for the most part, irreversible. If it is found that a well is indeed contaminating an aquifer or reemerging at the surface some distance away from the well, little can be done to correct the situation. Once the waste material is in the ground, it is unlikely that it can ever be extracted. Therefore, the potential for adverse environmental impact of deep well injection is quite serious.

Contract Disposal

Contract disposal--the purchase of disposal services--is not a disposal technique because it simply transfers the problem to another location and person. It is, however, often the least expensive disposal option. Because of stricter Federal regulations on hazardous materials, disposal services are becoming increasingly available and economically competitive with alternative schemes. For small industries, such disposal services often represent the best combination of economics, dependability, and convenience.

References

1. U.S. Environmental Protection Agency. Handbook for Monitoring Industrial Wastewater. EPA 625/6-73-002. NTIS No. Pb 259-146. Aug. 1973.

2. U.S. Environmental Protection Agency. Methods for Chemical Analysis of Water and Waste. EPA 16020-07-71. NTIS No. Pb 211-968. 1971.

3. Code of Federal Regulations. Title 40, Protection of Environment. Part 136, Guidelines Establishing Test Procedures for the Analysis of Pollutants. Chapter 1, Environmental Protection Agency.

4. American Public Health Association (APHA). Standard Methods for the Examination of Water and Wastewater (14th ed.). APHA, Washington DC, 1976.

5. American Society for Testing and Materials (ASTM). Annual Book of Standards: Part 31, Water. ASTM, Philadelphia PA, 1975.

6. W. H. Weiss. "Spill Prevention and Control: A Special Report." Pollution Engineering, $\underline{8}$(11):22-29, 1976.

7. U.S. Environmental Protection Agency. Development Document for Electroplating Point Source Category. EPA 440/1-78-085. Feb. 1978. (Available from Effluent Guidelines Division, Office of Water and Hazardous Materials, U.S. Environmental Protection Agency, Washington DC 20460)

8. D. M. Di Toro. "Statistical Design of Equalization Basins." Proceedings American Society of Chemical Engineers Journal of Environmental Engineering Division, 101(EE6):917, 1975.

9. V. Novotny and A. J. Englunde, Jr. Water Research, 8:325-332, 1974.

10. V. Novotny and R. M. Stein. "Equalization of Time Variable Waste Loads." Proceedings American Society of Chemical Engineers Journal of Environmental Engineering Division, 102(EE3):613-625, 1976.

11. P. Grover. "A Waste Stream Management System." Chemical Engineering Progress, 73(12):71-73, 1977.

12. R. S. Boynton. Chemistry and Technology of Lime and Limestone. John Wiley and Sons, New York, 1966.

13. W. A. Parsons. Chemical Treatment of Sewage and Industrial Wastes. National Lime Association, Washington DC, 1965.

14. R. H. Borgwardt. "Increasing Limestone Utilization FGD Scrubbers." Proceedings 68th Annual AIChE Meeting (BO-174). American Institute of Chemical Engineers, Los Angeles CA, Nov. 1975.

15. J. M. Smith. Chemical Engineering Kinetics (2nd ed.). McGraw-Hill, New York, 1970.

16. L. E. Gates, J. R. Morton, and P. L. Fondy. "Selecting Agitator Systems To Suspend Solids in Liquids." Chemical Engineering, 83(11):144-150, May 1976.

17. L. E. Gates, R. W. Hicks, and D. S. Dickey. "Application Guidelines for Turbine Agitators." Chemical Engineering, 83(26):165-170, Dec. 1976.

18. R. W. Hicks, J. R. Morton, and J. C. Fenic. "How To Design Agitators for Desired Process Response." Chemical Engineering, 83(9):102-110, Apr. 1976.

19. F. A. Holland and F. S. Chapman. Liquid Mixing and Processing in Stirred Tanks. Reinhold Publishing, New York, 1966.

20. W. L. McCabe and J. C. Smith. Unit Operations of Chemical Engineering (3rd ed.). McGraw-Hill, New York, 1976.

21. V. W. Uhl and J. B. Gray. Mixing: Theory and Practice (Vols. 1 and 2). Academic Press, New York, 1966.

22. R. L. Moore. Neutralization of Wastewater by pH Control. Instrument Society of America, Pittsburgh PA, 1978.

23. D. L. Hoyle. "Designing for pH Control." Chemical Engineering, 83(24): 121-126, 1976.

24. R. R. Rautzen, R. R. Corpstein, and D. S. Dickey. "How to Use Scale-Up Methods for Turbine Agitators." Chemical Engineering, 83(23):119-126, Oct 1976.

25. S. J. Khang and O. Levenspiel. "The Mixing-Rate Number for Agitator-Stirred Tanks." Chemical Engineering, 83(21):141-142, 1976.

26. P. M. Huck et al. "Optimization of Polymer Flocculation of Heavy Metal Hydroxides." Journal Water Pollution Control Federation, 49(12):2411-2418, 1977.

27. F. G. Shinskey. pH and pIon Control in Process and Waste Streams. John Wiley and Sons, New York, 1973.

28. R. W. Jones. "Graphical Analysis of Continuous Reactions." Chemical Engineering Progress, 47(1):46-48, 1951.

29. J. W. Patterson. Wastewater Treatment Technology. Ann Arbor Science Publishers, Inc., Ann Arbor MI, 1975.

30. M. J. Thomas and T. L. Theis. "Effects of Selected Ions in the Removal of Chrome III Hydroxide." Journal Water Pollution Control Federation, 48(8):2032-2045, 1976.

31. U.S. Environmental Protection Agency. "Processes for Heavy Metal Removal From Plating Wastewaters." First Annual Conference on Advanced Pollution Control for the Metal Finishing Industry. EPA 600/8-78-010. NTIS No. Pb 282-443. May 1978.

32. K. L. Rohrer. "Lime, $CaCl_2$ Bent Fluoride Wastewater." Water and Wastes Engineering, 11(11):66-68, 1974.

33. T. W. Cadmen and R. W. Dellinger. "Techniques for Removing Metals From Process Wastewater." Chemical Engineering, 81(8):79-85, 1974.

34. K. Jakobsen and R. Laska. "Advanced Treatment Methods for Electroplating Wastes." Pollution Engineering, 9(10):42-46, Oct. 1977.

35. K. H. Lanouette and E. G. Paulson. "Treatment for Heavy Metals in Wastewater." Pollution Engineering, 8(10):55-57, 1976.

36. R. H. Perry and C. H. Chilton. Chemical Engineers Handbook (5th ed.). McGraw-Hill, New York, 1973.

37. R. S. Boynton and K. A. Gutschick. Lime Handling Application and Storage. Bulletin 213. National Lime Association, Washington DC, 1971.

38. U.S. Environmental Protection Agency. User's Design Manual for the Treatment of Acid Mine Drainage, Draft Report. EPA Contract No. 68-03-2599. 1978.

39. "1979 Material Selector." Materials Engineering, 88(6), Dec. 1978.

40. National Association of Corrosion Engineers (NACE). Corrosion Data Survey (5th ed.). NACE, Houston TX, 1974.

41. Portland Cement Association (PCA). Effect of Various Substances on Concrete and Protective Treatments, Where Required. General Information Bulletin 3. PCA, Skokie IL, 1968.

42. J. Javetski. "Solving Corrosion Problems in Air Pollution Control Equipment." Power, 122(6):80-87, 1978.

43. U.S. Environmental Protection Agency. Estimating Costs for Water Treatment as a Function of Size and Treatment Efficiency. EPA 600/2-78-182. NTIS No. Pb 285-274. Aug. 1978.

44. U.S. Environmental Protection Agency. Controlling Pollution From the Manufacturing and Coating of Metal Products: Water Pollution Control. EPA 625/3-77-009. NTIS No. Pb 299-674. May 1977. (Prepared by Centec Consultants, Inc.)

45. Metcalf and Eddy, Incorporated. Wastewater Engineering: Treatment, Collection, Disposal. McGraw-Hill, New York, 1972.

46. I. V. Janerus and D. K. Lucas. "Settling and Thickening Metal Hydroxides with the Lamella Gravity Settler." Paper presented at the Pennsylvania Water Pollution Control Association Hershey Technical Conference, Hershey PA, June 15, 1977.

47. W. P. Talmadge and E. B. Fitch. "Determining Thickener Unit Areas." Industrial and Engineering Chemistry, 47:1, 1955.

48. Water Pollution Control Federation and American Society of Civil Engineers. Wastewater Treatment Plant Design, A Manual of Protection, MOP/8. Lancaster Press, Lancaster PA, 1977.

49. U.S. Environmental Protection Agency. Process Design Manual for Suspended Solids Removal. EPA 625/1-75-003a. NTIS No. Pb 259-147. Jan. 1975.

50. R. I. Dick. "Sludge Treatment." In W. J. Weber, Jr. (ed.), Physicochemical Processes for Water Quality Control. Wiley-Interscience, New York, 1972.

51. A. Hazen. "On Sedimentation." *Transactions of the American Society of Civil Engineers*, 53:45, 1904.

52. W. J. Katz and A. Geinopolis. "A Comparative Study of Hydraulic Characteristics of Two Types of Circular Solids Separation Basins." In J. McCabe and W. W. Eckenfelder, Jr. (eds.), *Biological Treatment of Sewage and Industrial Wastes* (Vol. 2). Reinhold Publishing, New York, 1958.

53. T. R. Camp. "Sedimentation and Design of Settling Tanks." *Transactions of the American Society of Civil Engineers*, 111:895, 1946.

54. J. R. Baylis. "Settling Basins for Coagulated Water." *Water and Sewage Works*, 78, 1931.

55. J. L. Cleasby, E. R. Baumann, and L. Schmid. *Comparison of Peripheral Feed and Center Feed Settling Tanks Using Models*. Iowa Engineering Experiment Station Report. Ames IA, Feb. 1962.

56. P. A. Vesilind. *Treatment and Disposal of Wastewater Sludges* (2nd ed.). Ann Arbor Science Publishers, Ann Arbor MI, 1979.

57. V. J. Jordan and C. H. Scherer. "Gravity Thickening Techniques at a Water Reclamation Plant." *Journal Water Pollution Control Federation*, 42(1):180-190, 1970.

58. U.S. Environmental Protection Agency. *The Reclamation of Metal Values From Metal Finishing Waste Treatment Sludges*. EPA 670/2-75-018. NTIS No. Pb 242-018. Apr. 1975.

59. R. I. Dick and B. Ewing. "Evaluation of Activated Sludge Thickening Theories." *American Society of Civil Engineers Journal Sanitary Engineering Division*, 94(SA4):9-30, 1967.

60. P. A. Vesilind. "Discussion of Dick and Ewing." American Society of Civil Engineers Journal Sanitary Engineering Division, 94(SA1):185-191, 1969.

61. G. L. Culp. Handbook of Sludge Handling Processes. Garland STPM Press, New York, 1979.

62. D. B. Chambers and W. R. T. Cottrell. "Flotation: Two Fresh Ways To Treat Effluents." Chemical Engineering, 83(16):95-98, 1976.

63. E. R. Vrablik. "Fundamental Principles of Dissolved Air Flotation of Industrial Wastes." Paper presented at the 14th Industrial Waste Conference, Purdue University, Lafayete IN, 1959.

64. W. W. Eckenfelder, Jr. Industrial Water Pollution Control. McGraw-Hill, New York, 1966.

65. V. Gulas et al. "Factors Affecting the Design of Dissolved Air Flotation Systems." Journal Water Pollution Control Federation, 50(7):1835-1840, 1978.

66. R. Braun. "Problems in the Removal of Inorganic Industrial Slurries." Proceedings of ALCHEMO Symposium. Chemical Publishing Company, 1974.

67. R. F. Wood and R. I. Dick. "Factors Influencing Batch Flotation Tests." Journal Water Pollution Control Federation, 45(2):304-315, 1973.

68. C. M. Ambler. "The Evaluation of Centrifuge Performance." Chemical Engineering Progress, 48(3):150-158, 1952.

69. P. A. Vesilind. "Characterization of Sludge for Centrifugal Dewatering." Filtration and Separation, Mar./Apr. 1977.

70. P. A. Vesilind. "Estimation of Sludge Centrifuge Performance." <u>American Society of Civil Engineers Journal Sanitary Engineering Division</u>, <u>96</u>(SA3):805-818, 1970.

71. R. C. Barkerville, A. M. Bruce, and M. C. Day. "Laboratory Technique for Predicting and Evaluating the Performance of a Filterbelt Press." <u>Filtration and Separation</u>, Sept./Oct. 1978.

72. U.S. Environmental Protection Agency. <u>Pressure Filtration of Waste Water Sludge With Ash Filter Aid</u>. EPA R2-73-231. NTIS No. Pb 223-535. June 1973.

73. U.S. Environmental Protection Agency. <u>Decision Makers Guide to Solid Waste Management</u>. EPA SW-500. NTIS No. Pb 258-266. 1976.

74. Radian Corporation. "Environmental Control Selection Methodology for a Coal Conversion Demonstration Facility." Radian Corporation, Houston TX, 1978.

Bibliography

Alspaugh, T. A. "Treating Dye Wastewater." *Textile Chemist and Colorist*, 5(11):255-260, 1973.

Beals, J. L. "Mechanics of Lime Slurries." *Proceedings: 37th International Water Conference, Engineers Society of Western Pennsylvania*. Pittsburgh PA, Oct. 1976.

Bituminous Coal Research, Inc. "Studies on Limestone Neutralization of Acid Mine Drainage." DAST-33, 14010 E 17 01/70, U.S. Department of the Interior, FWPCA, Washington DC, Jan. 1970.

Camp, T. R., and G. F. Conklin. "Rational Jar Test for Coagulation." *Journal New England Water Works Association*, 84(3):325-328, 1970.

Casto, L. V. "Practical Tips on Designing Turbine Mixer Systems." *Chemical Engineering*, 79(1):97-102, Jan. 10, 1972.

Chen, S. J., and A. R. Macdonald. "Motionless Mixers for Viscous Polymers." *Chemical Engineering*, 80(7):105-111, Mar. 19, 1973.

Cheremisinoff, P. N., I. Fideli, and N. P. Cheremisinoff. "Corrosion Resistance of Piping and Construction Materials." *Pollution Engineering*, 54(8):23-26, 1973.

Childs, C. W., P. S. Hallmon, and D. D. Perrin. "The Application of Digital Computers in Analytical Chemistry." *Talanta*, 16:1119, 1969.

Chow, V. T. *Open Channel Hydraulics*. McGraw-Hill, New York, 1959.

Clark, W. M. *Oxidation-Reduction Potentials of Organic Systems*. Robert E. Krieger Publishing, Huntington NY, 1972.

Corbett, E. W., O. W. Hargrove, and R. S. Merrill. *A Summary of the Effects of Important Chemical Variables Upon the Performance of Lime/Limestone Wet Scrubbing Systems*. Electric Power Research Institute (FP-639), Dec. 1977.

Culp, R. L., and G. L. Culp. *Advanced Wastewater Treatment*. Van Nostrand Reinhold, New York, 1971.

Dickey, D. S., and J. C. Fenic. "Dimensional Analysis for Fluid Agitation Systems." *Chemical Engineering*, $\underline{83}$(1):139-145, Jan. 1976.

Eckenfelder, W. W. *Water Quality Engineering for Practicing Engineers*. Barnes and Noble, New York, 1970.

Eden, G. E., and G. A. J. Truesdale. *Iron Steel Inst.* (London), $\underline{164}$:281-284, 1950.

Faust, S. D., and H. E. Orford. *Industrial Engineering Chemistry*, $\underline{50}$(10): 1537, 1958.

Fischer, M. C. "The Role of the Lamella Gravity Settler, A Compact Inclined Plate Clarifier in the Chemical Process Industry." Paper presented at the Symposium on Recent Advances in Sedimentation Practice Fundamentals at the 85th National A.I.Ch.E. Meeting, Philadelphia PA, June 1978.

Gates, L. E., and T. L. Henley. "How to Select the Optimum Turbine." *Chemical Engineering*, $\underline{82}$(26):110-114, Dec. 1975.

Gebauer, A. G. "Design of High Rate Settlers." *Journal of the Environmental Engineering Division, Proceedings of the American Society of Civil Engineers*, $\underline{100}$(EE5), Oct. 1974.

Huck, P. M., et al. "Optimization of Polymer Flocculation of Heavy Metal Hydroxides." Journal Water Pollution Control Federation, 49(12):2411-2418, 1977.

Judkins, J. F., Jr., and W. A. Parsons. Journal Water Pollution Control Federation, 9:1625, 1969.

Kostenbader, P. D., and G. F. Haines. "High-Density Sludge Treats Acid Mine Drainage." Coal Age, Sept. 1970.

LaMer, V. K., and T. W. Healy. "Adsorption-Flocculation Reactions of Macromolecules at the Liquid Solid Interface." Rev. Pure App. Chem., 13:112-132, 1963.

Lancy Laboratories. The Capabilities and Costs of Technology Associated With the Achievement of the Requirements and Goals of the Federal Water Pollution Control Act, as Amended, for the Metal Finishing Industry. National Commission on Water Quality, 1975.

LeVine, R. Y., and W. Rudolfs. Proceedings 7th Industrial Waste Conference. Purdue University, p. 305, 1952.

McCabe, W. L., and J. C. Smith. Unit Operations of Chemical Engineering. McGraw-Hill, New York, 1956.

Micheals, A. S. "Aggregation of Suspensions by Polyelectrolytes." Industrial and Engineering Chemistry, 46:1485-1490, 1954.

Morel, F., and J. J. Morgan. "A Numerical Solution for Solution of Chemical Equilibria in Aqueous Systems." In W. Stumm and J. J. Morgan (eds.), Aquatic Chemistry. Wiley-Interscience, New York, 1970.

Nancollas, G. N., and N. Purdie. "The Kinetics of Crystal Growth." Quarterly Review (London), 18:1, 1964.

O'Conner, D. J., and W. W. Eckenfelder. "Evaluation of Laboratory Settling Data for Process Design." In J. McCabe and W. W. Eckenfelder, Jr. (eds.), Biological Treatment of Sewage and Industrial Wastes, Vol. 2. Reinhold Publishing, New York, 1958.

Orgel, L. E. An Introduction to Transition Metal Chemistry: Liquid-Field Theory. John Wiley & Sons, New York, 1960.

Parker, N. R. "Mixing." Chemical Engineering, 165-220, June 8, 1964.

Parsons, W. A. Chemical Treatment of Sewage and Industrial Wastes. National Lime Association, Washington DC, 1965.

Patterson, J. W. Technology and Economics of Industrial Pollution Abatement. IIEQ Document No. 76/22. Illinois Institute for Environmental Quality, 1976.

Paulson, E. G. "Water Pollution Control Programs and Systems." In H. F. Lund (ed.), Industrial Pollution Control Handbook. McGraw-Hill, New York, 1971.

Perrin, D. D., and I. G. Sayce. "Computer Calculation of Equilibrium Concentrations in Mixtures of Metal-Ions and Complexing Species." Talanta, $\underline{14}$:833, 1967.

Perry, R. H. (ed.). Chemical Engineers Handbook (4th ed.). McGraw-Hill, New York, 1963.

Pick, R. I. "Thickening." In E. F. Gloyna and W. W. Eckenfelder, Jr. (eds.), Advances in Water Quality Improvement - Physical and Chemical Processes. University of Texas Press, Austin TX, p. 358, 1970.

Posselt, H. S., A. H. Reides, and W. J. Weber, Jr. "Coagulation of Hydrous Manganese Dioxide." Journal American Water Works Association, $\underline{60}$(1): 48-68, 1968.

Pulaski, J. C., Jr. "The Effect of Calcium Ions and pH on the Destabilization of a Dilute Kaolinite Suspension by Synthetic Polymers." In W. J. Weber (ed.), Physicochemical Processes for Wastewater Control. John Wiley and Sons, New York, 1972.

Rabosky, J. G., and D. L. Koraido. "Gaging and Sampling Industrial Wastewaters." Chemical Engineering, 80(1):111-120, 1973.

Ringbom, A. Complexation of Analytical Chemistry. Interscience Publishers, New York, 1963.

Ruehrwein, R. A., and D. W. Word. "Mechanism of Clay Aggregation by Polyelectrolytes." Soil Science, 73:485-492, 1952.

Schwartzenbach, G. Compleximetric Titrations (2nd ed.). Translated from German by H. Irving. Interscience, New York, 1956.

Shinskey, F. G. "Adaptive Feedback Applied to Feedforward pH Control." Proceedings Advances in Instrumentation, Part 1. Instrument Society of America, 1970.

Smith, J. H., III. "The Advantage of a Crowd for Acid Waste Liquors." Mining Engineering, 24(12):57-59, 1972.

Stumm, W., and C. R. O'Melia. "Stoichiometry of Coagulation." Journal of the American Water Works Association, 60:514-539, 1968.

Stumm, W., and J. J. Morgan. Aquatic Chemistry. John Wiley and Sons, New York, 1970.

Stumm, W., and J. J. Morgan. "Chemical Aspects of Coagulation." Journal of the American Water Works Association, 54(8):971-991, 1962.

Svanks, K., and K. S. Shumate. Factors Controlling Sludge Density During Acid Mine Drainage Neutralization. Report No. 392X. Ohio Water Resources Research Center, Columbus OH, Oct. 1973.

Urquhart, L. C. (ed.). Civil Engineering Handbook (4th ed.). McGraw-Hill, New York, 1959.

U.S. Environmental Protection Agency. Chemical Speciation of Contaminants in FGD Sludge and Wastewater. Interim Report, EPA Contract 68-03-2371, 1978. (prepared by SCS Engineers, Long Beach CA)

U.S. Environmental Protection Agency. Combination Limestone-Lime Neutralization of Ferrous Iron Acid Mine Drainage. EPA-600/2-78-002. NTIS No. Pb 280-169. 1978.

U.S. Environmental Protection Agency. Controlling Pollution from the Manufacturing and Coating of Metal Products. EPA-625/3-77-009. NTIS No. Pb 299-672. 1977.

U.S. Environmental Protection Agency. Evaluation of Problems Related to Scaling in Limestone Wet Scrubbing. EPA-R2-73-214. NTIS No. Pb 221-159. 1973.

U.S. Environmental Protection Agency. In Process Pollution Abatement. Vol. 1: Upgrading Metal-Finishing Facilities To Reduce Pollution. EPA 625/3-73-002. NTIS No. Pb 260-546. July 1973.

U.S. Environmental Protection Agency. Limestone and Lime Neutralization of Ferrous Iron Acid Mine Drainage. EPA-600/2-77-101. NTIS No. Pb 270-911. 1977.

U.S. Environmental Protection Agency. Process Design Manual for Suspended Solids Removal. EPA 625/1-75-003a. NTIS No. Pb 259-147. 1975.

Verwey, E. J. W., and J. Overbeek. *Theory of the Stability of Hyophobic Colloids.* Elsevier Publishing, 1948.

Willis, R. M. "Tubular Settlers--A Technical Review." *American Water Works Association Journal,* 70(6):331, 1978.

Wilmoth, R. C., and J. F. Martin. "Neutralization Options for Acid Mine Drainage Control." Paper presented at Fifth National Conference on Energy and Environment, Cincinnati OH, Nov. 1977.

Appendix—Sample Problems

EXAMPLE 1

Problem

Twelve days of sampling yielded the normalized values shown in Table A-1 for average daily flow rate and average daily pollutant loading. Project the design values for flow and pollutant loading that would not be expected to be exceeded more than 1.0 percent of the time.

TABLE A-1. NORMALIZED VALUES FOR AVERAGE DAILY WASTEWATER FLOW RATE AND POLLUTANT LOADING

Day	Average daily flow rate, F (l/s)	Average daily pollutant loading, P (mol/s)
1	34.0	45.0
2	35.0	60.0
3	32.0	24.0
4	34.0	60.0
5	23.2	34.6
6	13.9	59.0
7	40.0	98.0
8	26.0	72.0
9	47.2	57.7
10	32.0	55.0
11	15.9	62.0
12	34.5	105.0

Solution

The values of wastewater flow, F, and pollutant loading, P, are ranked in order of ascending magnitude in Table A-2. The probability of occurrence of values less than or equal to each data value is calculated as follows: for data point 1, probability of occurrence = 100/2n, n = 12; therefore, the probability of occurrence of data point 1 = 100/(2 X 12) = 4.2 percent. The interval between subsequent probability values = 100/n = 100/12 = 8.33 percent. The calculated probabilities for each data point are listed in Table A-2.

TABLE A-2. DATA FOR EXAMPLE 1

Rank	Wastewater flow, F (l/s)	Pollutant loading, P (mol/s)	Probability of occurrence of values less than or equal to F and P
1	13.9	24.0	4.2
2	15.9	34.6	12.5
3	23.2	45.0	20.8
4	26.0	55.0	29.2
5	32.0	57.7	37.5
6	32.0	59.0	45.8
7	34.0	60.0	54.2
8	34.0	60.0	62.5
9	34.5	62.0	70.8
10	35.0	72.0	79.1
11	40.0	98.0	87.5
12	47.2	105.0	95.8

In plotting the data, when a data point is duplicated, the highest probability value is used. For example, a flow rate of 32 l/s (8.5 gal/s) occurs twice in the data. This point is considered to have a probability of occurrence of 45.8 percent.

If the data are distributed log normally, a straight line can be drawn through the data points on log-normal probability graph paper. For purposes of illustration, it was assumed that the data are distributed log normally, and a least-squares fit of a straight line through the data point was calculated. The median flow rate and pollutant loading is 28.3 l/s (7.5 gal/s) and 56.7 mol/s, respectively.

The flow and pollutant loading values that would not be expected to be exceeded more than 1 percent of the time are 58.5 l/s (15.5 gal/s) and 162.0 mol/s, respectively (Figure A-1).

EXAMPLE 2

Calculate the minimal reservoir volume required to process the wastewater, having the characteristics described in Table A-3, at the average flow rate of the wastewater. The flow rate characteristics of the wastewater are presented in Figure A-2(a).

Flow Equalization--Graphical Procedure

Figure A-2(b) presents the graphical procedure for the evaluation of minimal storage volume for flow equalization, which begins with the construction of a plot of cumulative volume versus time for the flow relation. A mean flow line, A, is drawn between the initial and final points of the cumulative volume line. Another mean flow line, B, is drawn parallel to A and is tangent to the low point on the cumulative volume line. The minimal storage volume is determined from the graph as the greatest difference in ordinates between the translated mean flow line, B, and the cumulative volume line. The minimal storage volume represented by the dimension E_1, the difference between the cumulative volume line and the translated mean flow line, amounts to 1,060,000 l minus 740,000 l, or 320,000 l (84,544 gal).

The graphical procedure can be used to evaluate minimal storage volumes required for processing wastewater at selected rates above the mean flow--a common industrial waste practice. For example, if the flow relation were to be processed at 1.5 times the mean flow, a flow line with a slope of 1.5 times the mean flow slope would be constructed. The selected flow line, C, would be translated to a point of tangency to the cumulative volume relation at the low point that produces the maximal difference between the line, D,

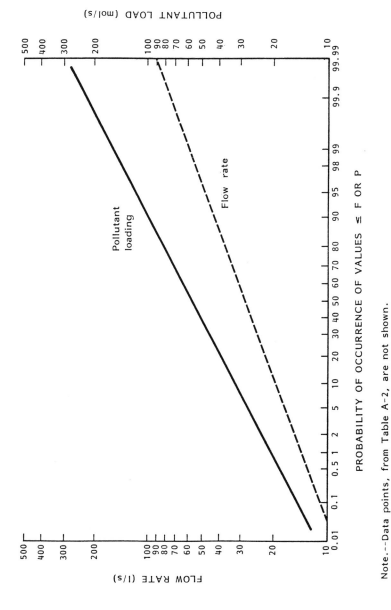

Figure A-1. Log normal distribution of data for Example 1.

and the curve. The construction is shown in Figure A-2(b), where the maximal difference is the dimension E_2, the difference between the cumulative volume line and the translated 1.5-times-mean-flow line. The value of E_2 is determined as 993,500 l minus 820,500 l, or 173,000 l (45,707 gal) of storage.

TABLE A-3. EFFLUENT SURVEY DATA FOR EXAMPLE 2

Time (h)	Flow rate (l/s)	Pollutant loading (mol/s)
1	20	2.0
2	18	1.8
3	10	0.05
4	5	0.03
5	5	0.01
6	15	0.03
7	17	1.70
8	30	6.00
9	82	164.0
10	32	128.0
11	37	72.0
12	25	100.0
13	15	6.00
14	15	6.30
15	20	6.00
16	25	60.00

Calculation of Minimal Reservoir Volume Required To Operate Continuously at Average Flow Rate

Over any period of time, the minimal reservoir volume required to maintain an average flow rate out of the reservoir is calculated by:

$$C_n = F_{av} t - \sum_{i=1}^{n} F_i t_i \qquad (A-1)$$

(a)

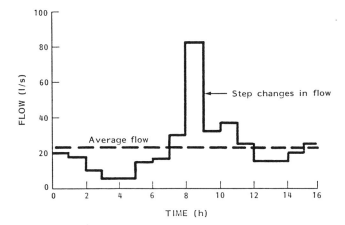

(b)

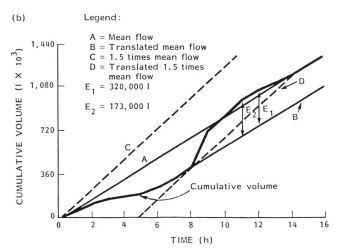

Figure A-2. Flow rate characteristics for two-shift industrial wastewater discharge: (a) average flow and step changes and (b) cumulative volume and minimal storage volumes.

where C_n = minimal reservoir volume (l)
       F_i = wastewater flow rate during time period, i (l/s)
       t_i = time period, i (s)
       F_{av} = average wastewater flow rate (l/s)
       n = number of time periods
       t = total time in n time periods (s)

Over the first 5 h, the cumulative volume of water discharged from the reservoir in excess of the volume of water that entered the reservoir is given by:

$$F_{av} = (20 + 18 + \ldots + 20 + 25)/16 = 23.2 \text{ l/s} \tag{A-2}$$

$$\begin{aligned} C_n &= 23.2 \text{ l/s} \times 5 \text{ h} \times 3{,}600 \text{ s/h} \\ &\quad - (20 \text{ l/s} + 18 \text{ l/s} + 10 \text{ l/s} + 5 \text{ l/s} + 5 \text{ l/s}) \\ &\quad \times 3{,}600 \text{ s/h} = 417{,}600 \text{ l} - 208{,}800 \text{ l} = 208{,}800 \text{ l} \end{aligned} \tag{A-3}$$

The minimal reservoir volume required for the system is determined by calculating the value of C_n for each time period. The minimal reservoir volume is then the difference between the maximum and minimum C_n.

In the sample problem, the maximal value of C_n occurs at hour 7 and the minimal value of C_n occurs at hour 12.

$$\begin{aligned} C_{12} &= (23.2 \text{ l/s} \times 12 \text{ h} \times 3{,}600 \text{ s/h}) - (20 \text{ l/s} + \ldots + 25 \text{ l/s}) \\ &\quad \times 3{,}600 \text{ s} = 1{,}002{,}240 \text{ l} - 1{,}065{,}600 \text{ l} = -63{,}360 \text{ l} \end{aligned}$$

$$\begin{aligned} C_7 &= (23.2 \text{ l/s} \times 7 \text{ h} \times 3{,}600 \text{ s/h}) - (20 \text{ l/s} + \ldots 17 \text{ l/s}) \\ &\quad \times 3{,}600 \text{ s} = 584{,}640 \text{ l} - 324{,}000 \text{ l} = 260{,}640 \text{ l} \end{aligned}$$

The minimal reservoir volume required is the difference between these values of C_{12} and C_7, or 324,000 l (85,600 gal).

EXAMPLE 3

Problem

A wastewater sample was taken from a plant and neutralized in the laboratory with a 12.3 $\underline{N}$ Ca(OH)$_2$ solution. The experimental data are tabulated in Table A-4, and Figure A-3 presents the titration curve. Calculate the concentration of residual reagent for each time increment and the reaction rate equation.

TABLE A-4. TITRATION AND KINETIC DATA FOR EXAMPLE 3

Time (min)	pH	Titration (ml/l)
0	1.33	0
0.25	1.73	9.15
0.50	1.96	10.68
0.75	2.18	11.75
1.00	2.28	12.03
1.50	2.56	12.63
2.00	2.75	12.77
2.50	2.96	12.97
3.00	3.22	13.10
3.50	3.52	13.16
4.00	3.96	13.23
4.50	4.40	13.27
5.00	5.50	13.33
6.00	6.28	13.37
7.00	7.32	13.39
8.00	8.08	13.43
9.00	8.50	13.46
10.00	8.88	13.48
11.00	8.93	13.49

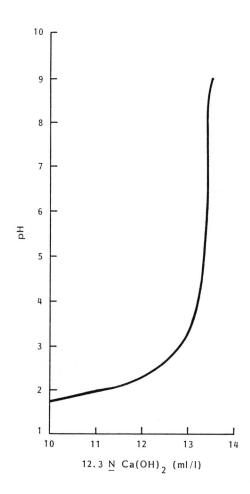

Figure A-3. Section of titration curve for neutralization of process waste with slurry of high calcium in Example 3.

Solutions

Calculation of Residual Reagent Concentration[b]

The following values are given:

- pH after 0.25 min = 1.73
- volume of solution required to titrate to a pH of 1.73 = 9.15 ml/l
- initial dosage of reagent = 13.5 ml
- residual reagent after 0.25 min = 4.35 ml
- molecular weight of $Ca(OH)_2$ = 74 g/g-mol
- equivalent weight of $Ca(OH)_2$ = 74/2 = 37 g-eq/g-mol

The residual reagent after 0.25 min is calculated as follows:

$$4.35 \text{ ml/l} \times 12.3 \underline{N} \times 1/1{,}000 \times 37/74 = 0.02675 \text{ g-mol/l}$$

$$0.02675 \text{ g-mol/l} \times 1{,}000 \text{ l/l} = 26.75 \text{ g-mol}/1{,}000 \text{ l}$$

The preceding calculation is repeated for each time increment, as shown in Table A-4. Residual reagent versus time and residual reagent versus pH are illustrated in Figures 16 and A-4, respectively.

Determination of Reaction Rate Equation

Differential Analysis. From the data in Table A-4 of residual reagent and time, it is possible to determine the reaction rate equation to describe the disappearance of B. These data are plotted in Figure 14. The slope at any point on the curve represents the reaction rate. The slope of a curve, however, cannot be calculated directly. Where a straight line is tangent to the desired curve, at the point of contact, the two lines have the same slope.

---

[b] Residual reagent is calculated in gram-moles per 1,000 liters of wastewater (g-mol/1,000 l).

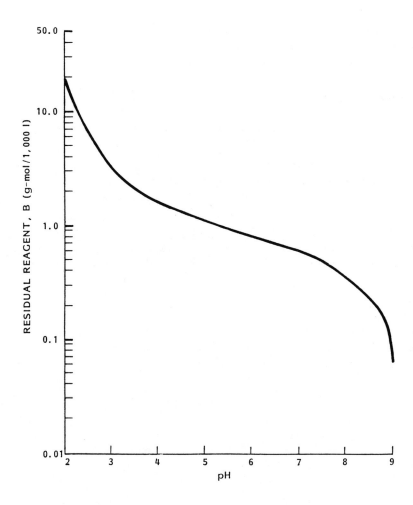

Figure A-4. Semilog plot of residual reagent, B, versus pH for batch reactor.

The algebraic equation for the desired neutralization curve is unknown; however, the equation of the line tangent to the neutralization curve can be found. Neutralization reactions can often be described by Equation 4 in Section 3. Taking the natural logarithm of this equation produces a straight line on semilog paper, which is described by the following equation:

$$\ln(y) = \ln(a) + x [\ln(b)] \qquad (A-4)$$

The corresponding equation for this line on linear paper is $y = a \, X \, b^x$. The slope of this curve at any point is the derivative:

$$dy/dx = (a \, X \, b^x)[\ln(b)] \qquad (A-5)$$

Thus, at the point of contact, we have the slope of the line and the slope of the curve at the point.

Care must be taken when drawing the line tangent to the curve. (An approximation of the slope can be taken graphically by measuring the smallest x and y increments possible to read accurately; however, this method is not recommended.) A line is drawn tangent to the curve in Figure 14 at B = 9. The equation of this line is:

$$\ln(B) = \ln(a) + t[\ln(b)] \qquad (A-6)$$

By substituting values of B and t from two points on the tangent line in the equation, two simultaneous equations are obtained from which the values of parameters a and b can be determined:

$$\ln(30) = \ln(a) + 0 \qquad (A-7)$$

at B = 30, t = 0

$$\ln(0.1) = \ln(a) + 4.63 \ln(b) \qquad (A-8)$$

at B = 0.1, t = 4.63

By simultaneous solution of the two equations, it is determined that $a = 30$ and $b = 0.292$.

Therefore, the slope of the curve at $B = 9$ and $t = 1$ is $dB/dt = 30 \times (0.292)^{(t)} \ln (0.292) = -10.78$.

This procedure is repeated for several values of B. The results of these calculations are tabulated in Table A-5 and illustrated in Figure A-5.

TABLE A-5. TABULATED RESULTS FOR DIFFERENTIAL ANALYSIS METHOD

$-dB/dt$ (g-mol/1,000 l-min)	B (g-mol/1,000 l)	Log (dB/dt)	Log B
224.000	83.000	2.350	1.919
25.000	17.000	1.398	1.230
10.780	9.000	1.033	0.954
1.410	2.350	0.149	0.371
0.537	1.040	-0.270	0.017
0.274	0.550	-0.562	-0.260
0.138	0.245	-0.860	-0.610
0.080	0.060	-1.097	-1.222

Drawing the lines tangent to various values of B is difficult because slight movements can give slopes of varying value. The line selected by the procedure might be somewhat different than selecting another that would give different values for the slope. This should not affect the overall procedure significantly, but it should be apparent that slopes could differ somewhat for this reason.

If the rate equation is of the form $dB/dt = -kB^n$, the data should plot as a straight line on log-coordinate graph paper, except when $n = 1$. By expressing the data in logarithmic form, as shown in Table A-5, and plotting these numbers on rectilinear paper, the "best" straight line through the data can be drawn by a regression analysis of the data points (Figure A-6). Note that the linearity pertains to the reactive fraction of the lime and not to the impurities. The following equation results:

$$y = -0.080 + 1.158x \tag{A-9}$$

where $y = \log dB/dt$
$x = \log B$
$\log k = -0.080$
$n = 1.158$

Solving for k:

$$k = 10^{-0.080} = 0.832$$

Therefore, the rate equation is:

$$dB/dt = -0.832 \, B^{1.158} \quad \text{(B is disappearing)} \tag{A-10}$$

Statistical Analysis. From the original data and graph of residual reagent, B, versus time, t, a multiple regression computer program was employed to fit the best curve. The resulting curve with a correlation coefficient of 0.9981 was hyperbolic with the general form of:

$$y = a + b/t \tag{A-11}$$

The regression gives the following curve:

$$B = -0.844 + 10/t \tag{A-12}$$

where B = residual reagent (g-mol)
t = time (min)

Differentiating Equation 12 to determine the rate dB/dt yields

$$dB/dt = -10/t^2$$

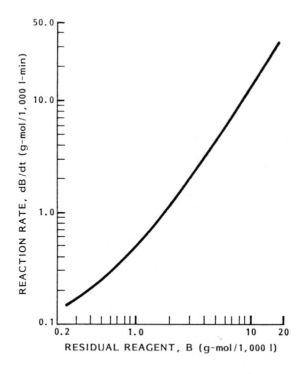

Figure A-5. Logarithmic plot of reaction rate versus residual reagent, B, for batch reactor.

Eliminating t from the rate equation by substitution from the original equation yields

$$t = 10/(B + 0.844)$$

$$r = dB/dt = -(B + 0.844)^2/10$$

The rate equation is now in terms of B.

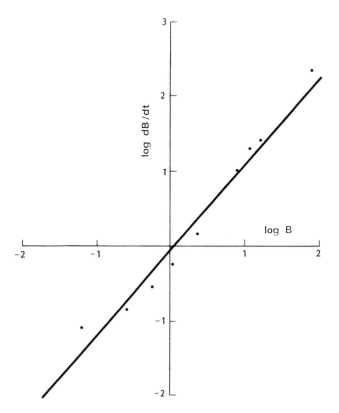

Note.--Inactive impurities in reagent are disregarded in this plot.

Figure A-6. Interpretation of logarithmic plot of reaction rate versus residual reagent, B, for analysis by the differential method.

Once a rate relation or equation has been determined, the design of a reactor system is possible through the use of the design equations described in Section 4 of this manual.

EXAMPLE 4

The following calculations use the reaction rate equation developed by the statistical technique to design staged reactors for the wastewater in Example 3. Subscript numbers are used to designate inlet or outlet concentration as well as reactor stage. The first subscript can be either 1 or 2 to designate inlet or outlet concentration, respectively, whereas the second subscript number refers to the reactor stage. For example, B_{21} is the <u>outlet</u> concentration of B for the <u>first</u> stage.

$$B_{21} = r\theta + B_{11}$$

If $B_{11} = 83$ and $\theta = 10$ min:

$$r = -(B_{21} + 0.844)^2/10$$

$$B_{21} = -[(B_{21} + 0.844)^2/10] \times 10 + 83$$

Simplifying the above expression gives the quadratic equation:

$$B_{21}^2 + 2.69B_{21} - 82.288 = 0$$

Solving for B_{21} will give the value for the concentration of residual reagent in the exit stream from the first stage with a 10-min retention time (B_{21}). From Figure A-4, $B_{21} = 7.83$ and the corresponding pH = 2.38.

For Stage 2, $B_{12} = 7.83$ and

$$B_{22}^2 + 2.69B_{22} - 6.98 = 0$$

where $B_{22} = 1.62$
pH = 4.0

For Stage 3, $B_{13} = 1.62$ and

$$B_{23}^2 + 2.69B_{23} - 0.90 = 0$$

where $B_{23} = 0.30$
      $pH = 8.25$

This brings the effluent within the desired range.

EXAMPLE 5

Using the Jones method, the values of $-dB/dt$ versus B are graphed on log-log paper, as shown in Figure A-7.

Starting at the initial value of $B = B_1 = 83$, a graph is drawn of the equation:

$$dB/dt = 1/\theta \, (B_1 - B_2) \qquad (A-13)$$

where $(1/\theta \times B_1)$ = constant
      θ = 10-min retention time

On linear paper, this equation could be drawn as a straight line. On log-log paper, several points must be plotted and the corresponding curve drawn between the points, as shown in Figure A-7. The point at which this curve intersects the $-dB/dt$ curve gives the value of the exit concentration of B in the first stage for a 10-min retention time. The equation is drawn as follows: Calculate $1/\theta \, (B_1)$ at $dB/dt = 0.01$ and $B_2 = 83$.

$$1/\theta \, (B_1) = 1/\theta \, (B_2) + dB/dt$$

$$1/\theta \, (B_1) = 0.1(83) + 0.01 = 8.31$$

The value of $(1/\theta \times B_1)$ remains constant, and so several values of B_2 are selected to calculate dB/dt. The results of these calculations are given in Table A-6.

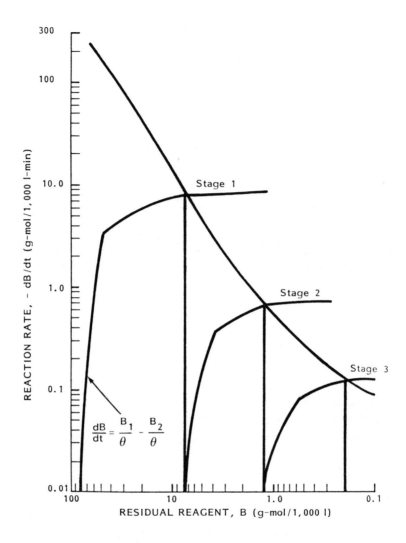

Figure A-7. Graphical projection of batch reactor results to continuous flow reactor performance by modified Jones method.

TABLE A-6. VALUES OF B_2 and dB/dt

B_2 (g-mol/1,000 l)	dB/dt (g-mol/1,000 l-min)
83	0.01
50	3.31
30	5.31
10	7.31
6	7.71
5	7.81

The best curve through these points is drawn on the graph. The curve intersects the $-dB/dt$ curve at the concentration of B_2 in the first reactor stage. Thus, after the first stage, $B_2 = 7.6$ and pH = 2.40.

For the second stage, the procedure is repeated starting the curve at $B_2 = 7.6$ and $dB/dt = 0.01$, with θ again equal to 10 min. At this point, $1/\theta(B_1) = 0.1(7.6) + 0.01 = 0.77$:

$$dB/dt = 1/\theta(B_1) - 1/\theta(B_2) = 0.77 - 0.1B_2$$

Plotting these points forms a curve that intersects the dB/dt curve at $B_2 = 1.25$. After the second stage, $B_2 = 1.25$ and pH = 4.66. Performing the same procedures for a third stage results in $B_2 = 0.2$ and pH = 8.65. Three stages are required to complete the neutralization.

EXAMPLE 6

Given

A wastewater neutralization system will be designed to accommodate a mean flow of 28.32 l/s (7.48 gal/s), which is projected to range from a minimum of 9.35 l/s (2.47 gal/s) to a maximum of 47.3 l/s (12.5 gal/s). The pH is to be controlled at 6.5 in the reactor. Survey results indicate that the acid concentration will range from 0.03 $\underline{N}$ to 0.318 $\underline{N}$ in a period of 1.8 min. The maximal pollutant loading (acidity mass flow) is projected to be 16.96 g-eq/s. The wastewater temperature is 25° C (77° F) and the titration curve is shown in Figure A-8. Laboratory studies in a 40-1 (10.6-gal) standard reactor indicate the reaction rate relation to be $-dB/dt = 1.5B^{1.5}$ at a reagent dosage of 2 percent above that required for a 6.5 pH. Agitation was by a pitched blade turbine running at 500 r/min, and the density of the mixture was 1,105 kg/m^3 (70 lb/ft^3). The viscosity was determined as 0.0062 kg/m-s (0.0042 lb/ft-s).

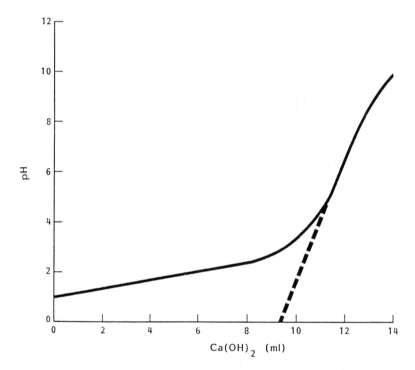

Figure A-8. Titration of 200 ml of acid wastewater with 5.43 $\underline{N}$ Ca(OH)$_2$ lime slurry.

Required

The following steps must be performed to solve the problem:

- Determine prototype reactor volume and dimensions.
- Scale up for comparable solids suspension and check blend time for A = 0.01.
- Estimate power requirement.
- Estimate reactor dead time.

Reactor Volume

Take the design situation as the projected maximal pollutant loading in the maximal flow. From Figure A-8, note that with 2 percent alkali overdose, the effluent pH may drift to 6.6.

$$C_i = \frac{\text{pollutant loading}}{\text{design flow}} = \frac{16.96 \text{ g-eq/s}}{47.3 \text{ l/s}} = 0.359 \underline{N}$$

From Figure A-8, a pH of 6.5 requires 12/13 of reagent to reach a pH of 7.0. Take material balance at steady flow, letting:

F = wastewater flow = 47.3 l/s (or 12.5 gal/s)
R = reagent feed = (0.359/0.176)(12.0 ml/400 ml)47.3 l/s = 2.89 l/s
   (or 0.76 gal/s)
C_e = effluent acid concentration = (1/13)(0.359) = 0.028 $\underline{N}$

$$F_i C_i = (F + R)C_e + kVB_e^n + 0 \qquad (A\text{-}14)$$

47.3 × 0.359 = (47.3 + 2.9)0.028 + 1.5V(12/13 × 0.359 × 0.02)$^{1.5}$

where V = 19,240 l (or 5,083 gal)

Apply design factor of 10 percent for contingency primarily for variation in commercial grade reagents. Trial reactor volume, V = 1.1 X 19,240 l = 21,160 l (or 5,590 gal). Retention time, θ = V/(F + R) = 21,160/(47.3 + 2.9) = 421.5 s or 7.03 min. Use depth equal to diameter (i.e., H = T):

$$V = 21,160 \text{ l} = \pi/4 \ (T^2H)$$

T = H = 3.0 m (or 9.8 ft) = prototype reactor diameter

Experimental reactor volume = 40 l (or 10.6 gal).

$$V = 40 \text{ l} = (\pi/4) \ (T^2H)$$

T = H = 0.37 m (or 1.217 ft) = model reactor diameter

Scale-Up

To scale up for comparable solids suspension, use exponent m = 0.75 from Equation 27.

$$n_2/n_1 = (D_1/D_2)^m = (T_1/T_2)^m \text{ for } D = T/3$$

where D = propeller diameter = 1/3 tank diameter, T

$$n_2 = 500 \text{ r/min } (0.371/3.0)^{0.75} = 104.3 \text{ r/min} = 1.74 \text{ r/s}$$

For blend time, take A = 0.01, and estimate N_{Re} for prototype and model. From Equation 24:

$$N_{Re} = \frac{\rho n D^2}{\mu} = \frac{\rho n (T/3)^2}{\mu}$$

$$N_{Re} = \frac{(1,105 \text{ kg/m}^3)(1.74 \text{ r/s})(3.0 \text{ m/3})^2}{0.0062 \text{ kg/m-s}} = 310,000 \text{ (prototype, turbulent)}$$

$$N_{Re} = \frac{(1,105 \text{ kg/m}^3)(500 \text{ r/min/60})(0.371 \text{ m/3})^2}{0.0062 \text{ kg/m-s}} = 22,700 \text{ (model, turbulent)}$$

From Figure 17, N_{MR} = 0.58 for prototype and model:

$$K = \frac{n(D/T)^{2.3}}{N_{MR}} = \frac{1.74 \text{ r/s } (1/3)^{2.3}}{0.58} = 0.240/\text{s (prototype)}$$

$$K = \frac{(500 \text{ r/min}/60)(1/3)^{2.3}}{0.58} = 1.15/\text{s (model)}$$

$$t_b = (1/K)(\ln)(2/A) = 1/(0.240/\text{s}) \times \ln(2/0.01) = 22.1 \text{ s (prototype)}$$

$$t_b = 1/(1.15/\text{s}) \times \ln(2/0.01) = 4.61 \text{ s (model)}$$

$$\Delta t_b = 22.1 - 4.61 = 17.5 \text{ s}$$

The value of $\Delta t_b/\theta$ = 17.5/421.5 = 0.042, or 0.01 percent, which is less than the 10-percent contingency allowance.

Reactor Power Requirement

From Figure 15, N_{Re} = 308,000 and N_p = 1.23. From Equation 23,

$$N_p = \frac{Pg}{\rho n^3 D^5} = \frac{P(9.81 \text{ m/s}^2)}{(1,105 \text{ kg/m}^3)(1.74 \text{ r/s})^3 (3.0 \text{ m/3})^5} = 1.23$$

Solving:

$$P = 730.0 \text{ kg-m/s} = 7.16 \text{ kW (or 9.6 hp)}$$

Reactor Dead Time

Reactor volume is 21,160 l (5,590 gal). The minimal number of kilowatts per 1,000 liters is:

$$3.959 \text{ V}^{-0.283} = 3.959(21,160)^{-0.283} = 0.236$$

$$\frac{7.16 \text{ kW}}{21,160 \text{ l}} \times 1,000 = 0.338 \text{ kW}/1,000 \text{ l}$$

Because 0.338 kW/1,000 l (1.72 hp/1,000 gal) is greater than the minimum of 0.236 kW/1,000 l (1.20 hp/1,000 gal), it is an acceptable value.

From Equation 25, with $N_Q = 0.75$:

$$N_Q = Q/nD^3 = Q/(1.74 \text{ r/s})(3.0 \text{ m/3})^3 = 0.75$$

where Q = 1,300 l/s

Tank turnovers, therefore, equal (1,300 l/s)(60 s/min)/21,160 l = 3.70/min.

From Figure 16, dead time equals 10 s. Thus, $T_d/\theta = 10/421.5 = 0.02$, which is acceptable because it is less than 0.05.

Discussion

The design situation entails a volumetric scale-up of 21,160/40 = 529. The integrity of the scale-up would be improved by the experimental evaluation of exponent m in Equation 27. The value of m can be determined from analysis of experimental results obtained from reactors of different size. It should be recognized that the prototype power requirement is extremely sensitive to the impeller diameter (fifth power) and the impeller speed (third power).

Variations in wastewater and reagent characteristics also warrant consideration. The soluble alkalies (ammonia, sodium hydroxide, and sodium carbonate) are consistent in neutralization characteristics. High calcium lime usually has consistent reaction characteristics--especially if obtained from the same supplier. Dolomitic lime and carbide lime are more subject to variation in reaction characteristics, especially when obtained from different suppliers. Limestone is a highly variable reagent that requires careful quality control during purchasing to obtain a supply with consistent reaction characteristics.

Scale-up on the basis of solids suspension is advocated inasmuch as reaction rate is frequently a weak function of agitation level when all solids are in suspension. Volumetric compensation for differences in blend time between model and prototype is provided to give comparable time for the instrumented feed system to act on disturbances.

EXAMPLE 7

Given

Neutralization of 1 liter (0.26 gal) of waste requires 10 grams (0.35 oz) of reagent added at a rate of 0.5 g/l/min (0.06 oz/gal/min) to achieve the proper sludge or precipitate properties. This requirement corresponds to a selected recycle sludge rate. Thus, on a batch basis, 20-min retention could be allotted to control both crystallization and pH, assuming that the reaction reaches equilibrium at the time reagent addition is complete. If equilibrium is not reached, additional retention time would be provided. For an effluent flow rate of 100 l/min (26 gal/min), size the reactors for a continuous neutralization system.

Solution

One hundred grams (3.5 oz) of reagent added at 0.5 g/l/min (0.06 oz/gal/min) is required for pH control.

The first step in the design procedure is to assume a trial number of reactor stages in the system. If feedback pH control will be used, the rate of reagent addition to all but the last reactor can usually be assumed to be governed by the laboratory-determined reagent feed rate that optimizes sludge properties. That is, if five reactors in series are selected for the system, the design will put the first four reactors under the control of the predetermined rate addition, and the fifth reactor will be controlled by the objective pH set point and the reaction chemistry. The objective of the analysis that follows will be to outline a procedure for establishing the set points for the first four reactors. Some acid/alkali systems are unsuited to feedback pH control because of high buffer capacity. Control by automatic titration can be adapted to such systems.

With the titration curve, the reagent dose, the reaction rate function, the return sludge dosage, and the optimal rate of reagent addition established from laboratory studies, the first step in the analysis is to construct a plot of log dB/dt versus log B. The transition point between control by reagent addition and control by final pH set point can be determined from the following relation:

$$\text{Log } R = \text{Log } (dB/dt) \qquad (A-15)$$

where R = rate of reagent addition for optimal sludge
  dB/dt = chemical reaction rate from batch process experiment (g/l-min)
  B = residual reagent (g/l)

The transition point is determined from input of the value of log R into the plot of log dB/dt versus log B. For illustration purposes, the transition point can be taken as B = 0.02 g/l (0.003 oz/gal). The transition value of B is often small and can be used as an approximation of B_e between stages with this design procedure. The residual reagent to be reacted in the first four reactors amounts to 10 - 0.02 = 9.98 g/l (1.28 oz/gal). It is convenient to react 2.48 g/l (0.32 oz/gal) in the first stage and 2.5 g/l (0.32 oz/gal) in Stages 2, 3, and 4.

The volume of the first reactor is obtained from the material balance relation:

$$FB_i = FB_e + RV \qquad (A-16)$$

where F = total flow = 100 l/min (26 gal/min)
  B_i = influent reagent concentration = 2.5 g/l (0.32 oz/gal)
  B_e = effluent residual reagent concentration = 0.02 g/l (0.003 oz/gal)[c]
  R = reagent addition rate = 0.5 g/l-min (0.06 oz/gal-min)
  V = reactor volume (l)

---

[c] Appropriate value selected by the designer.

Solving Equation A-16 for V yields:

$$V = (100)(2.5 - 0.2)/0.5 = 496 \text{ l (or 131 gal)}$$

The pH set point for the first stage reactor is taken from the titration curve at a value of B = 2.48 g/l (0.32 oz/gal). The value of input reagent concentration, B_i, becomes (2.5 + 0.02) g/l (0.32 oz/gal) for the second, third, and fourth reactor stages. Therefore:

$$V = (100)(2.52 - 0.02)/0.5 = 500 \text{ l (132 gal)}$$

The pH set points for the reactor stages would be taken from the titration curve corresponding to the following values of B:

- B = 2.48 + 2.50 = 4.98 g/l (0.64 oz/gal)--Stage 2
- B = 2.48 + 2 X 2.5 = 7.48 g/l (0.96 oz/gal)--Stage 3
- B = 2.48 + 3 X 2.5 = 9.98 g/l (1.28 oz/gal)--Stage 4

For illustration, Stage 5 will operate with a feed, B_i, of 0.02 g/l (0.003 oz/gal) and a neutral effluent, B_e, of 4.0×10^{-5} g/l (5.12×10^{-6} oz/gal). The reaction will be governed by a rate relation such as $dB/dt = kB^n$. The steady-state material balance relation becomes:

$$FB_i = FB_e + kB_e^n V$$

The pH set point corresponds to a B_e of 4×10^{-5}.

The five-stage reactor system would be expected to provide excellent control in the absence of substantial flow fluctuations but probably would be less economical than a reactor system with fewer stages (e.g., three stages). With small tanks, extra stages involve extra instrumentation and equipment that add to the comparative cost. The feasibility of a system with fewer stages should be evaluated as the next step in the design. If different sized stages are desired, the objective may be attained by assignment of different conversion allotments to the respective stages.

EXAMPLE 8

Given

Alternative neutralization systems are designed to accommodate an acid waste contributing an average pollutant load of 4.956 g-eq/s and a maximal pollutant load of 16.96 g-eq/s. One system uses 50 percent sodium hydroxide (NaOH) for neutralization and has a reactor volume of 9,930 l (2,624 gal). The alternative system uses bulk, high calcium hydrated lime [$Ca(OH)_2$] slurry for neutralization and has a reactor volume of 21,310 l (5,630 gal).

Required

Estimate the capital cost of each system and make a first order annualized cost comparison of the alternatives.

Estimate capital cost of mixed reactor:

- NaOH--from Figure 24 @ 9,930 l (2,624 gal); cost = $18,800
- Lime--from Figure 24 @ 21,310 l (5,630 gal); cost = $23,200

Estimate capital cost of feed systems plus pH control system at feed rate of 16.96 g-eq/s.

$$NaOH = \frac{16.96 \text{ g-eq/s} \times 40 \text{ g/g-eq}}{1,000 \text{ g/kg}} \times 3,600 \text{ s/h} = 2,442 \text{ kg/h (or 5,384 lb/h)}$$

From Figure 25, the capital cost is $317,000.

For the lime system, assume 15 percent slurry as $Ca(OH)_2$ and take the specific gravity of a $Ca(OH)_2$ particle as 2.5.

$$\text{Slurry} = 15 \text{ g (or 0.5 oz) } Ca(OH)_2 + 85 \text{ g (or 3 oz) } H_2O$$

$$\text{Slurry volume} = 15/2.5 + 85 = 91.0 \text{ ml (or 3.1 oz)}$$

$$\text{Slurry} = \frac{15 \times 1{,}000/91}{74/2} = 4.455 \text{ } \underline{N}$$

$$\text{Maximal slurry flow} = \frac{16.96 \text{ g-eq/s}}{4{,}455 \text{ g-eq/l}} = 3.81 \text{ l/s (or 1 gal/s)}$$

$$16.96 \text{ g-eq/s} = \frac{16.96 \times 3{,}600 \text{ s/h}}{1{,}000 \text{ g/kg}} \times 74/2 = 2{,}259 \text{ kg/h (or 4,976 lb/h)}$$

From Figure 27, the cost is $550,000 (see Table A-7).

Calculate Annualized Costs

For liquid sodium hydroxide, assume that costs for operation (exclusive) of reagents) and maintenance are similar to those for liquid alum, as given in Figure 26. Assume that the costs for the feed of lime are 1.2 times those given in Figure 28 for dry alum.

For 350 days on stream, the average reagent demand of 4.956 g-eq/s is equivalent to 5,995 metric tons/yr (6,608 short tons/yr) of NaOH and 5,717 metric tons/yr (6,302 short tons/yr) of $Ca(OH)_2$:

$$\frac{4.956 \text{ g-eq/s} \times 40 \text{ g/g-eq} \times 3{,}600 \text{ s/h} \times 24 \text{ h/d} \times 350 \text{ d/yr}}{1{,}000 \text{ g/kg} \times 1{,}000 \text{ kg/metric ton}} = 5{,}995 \text{ metric tons per year (NaOH)}$$

$$\frac{4.956 \times 74/2 \times 3{,}600 \times 24 \times 350}{1{,}000 \times 1{,}000 \times (0.97 \text{ purity})} = 5{,}717 \text{ metric tons per year } [Ca(OH)_2]$$

The following rates are assumed: NaOH at $180/metric ton ($163/short ton), $Ca(OH)_2$ at $37.50/metric ton ($34.02/short ton), and shipping charges at $12/metric ton ($10.89/short ton) of NaOH and $8/metric ton ($7.26/short ton) of $Ca(OH)_2$. As shown in Table A-8, the annual cost of NaOH = 5,995 X (180 + 12) = $1,151,040, and the annual cost of $Ca(OH)_2$ = 5,717 X (37.50 + 8) = $260,124.

TABLE A-7. CAPITAL COST ESTIMATE

Unit Process	Cost ($)	
	NaOH	Ca(OH)$_2$
Reactor	18,800	23,200
Reagent, storage, feed, and controls	317,000	550,000
Total	335,800	573,200
Sitework, interface piping, and roads @ 0.05 X 336,000	16,800	16,800
Subsurface anomalies and standby power	(a)	(a)
Total	352,600	590,000
General contractor overhead and profit @ 15%	52,890	88,500
Total	405,490	678,500
Engineering @ 10%	40,550	67,850
Total	446,040	746,350
Land	(a)	(a)
Legal and administrative (Figure 39)	11,000	15,000
Interest during construction @ 8% (Figure 40)	36,560	60,910
Total capital cost [excluding (a)]	493,600	822,260

[a] As applicable to specific situation.

SOURCE: Procedure adapted from more detailed presentation in U.S. Environmental Protection Agency, *Estimating Costs for Water Treatment as a Function of Size and Treatment Efficiency*, EPA 600/2-78-182, NTIS No. Pb 285-274, Aug. 1978.

TABLE A-8. ANNUALIZED COSTS

Cost Component	Cost ($)	
	NaOH	Ca(OH)$_2$
Capital cost @ 20 percent	94,808	159,470
Reagent cost	1,151,040	260,124
Operation and maintenance	4,700[a]	20,000[b]
Instrumentation maintenance	22,000	30,000
Total	1,272,548	469,594

[a] Similar to cost for liquid alum (Figure 26).

[b] 1.2 times cost for dry alum (Figure 28).

EXAMPLE 9

Given

Wastewater flows through a rectangular sedimentation basin--measuring 10 m (33 ft) wide, 30 m (98 ft) long, and 3 m (9.8 ft) deep--at a rate of 100 million l/d (26 million gal/d). The settling velocity distribution of particles and the overall particle removal efficiency are given in Table A-9.

Solution

The settling velocity of the particles that will be completely removed is calculated as follows:

$$V_t = \frac{Q}{A} = \frac{100 \times 10^6 \text{ l/d} \times (1/1,440)(\text{d/min}) \times 100 \text{ cm}^3/\text{l}}{(10 \times 30)\text{m}^2 \times 10,000 \text{ cm}^2/\text{m}^2}$$

$$= 2.3 \text{ cm/min (or 9 in/min)}$$

Laboratory tests are useful in constructing a settling velocity analysis curve to determine the removal efficiency for a given settling time. Although sieve analyses and hydrometer tests may be used to obtain a particle size distribution from which a particle velocity distribution can be determined, settling column tests are conducted more often. Reference 45 contains a more detailed discussion of such procedures.

TABLE A-9. SETTLING VELOCITY DISTRIBUTION AND PARTICLE REMOVAL EFFICIENCY

Column 1: Particles (%)	Column 2: Settling velocity (cm/min)	Column 3: Particles removed (%) V_s/V_t	Column 4: Overall particle removal efficiency (%) [(Column 1 X Column 3)/100]
5	5	22	1.1
8	8	35	2.8
10	10	43	4.3
12	12	52	6.2
15	15	65	9.8
15	20	87	13.1
12	23	100	12.0
10	25	100	10.0
8	28	100	8.0
5	>30	100	5.0
			72.3

Other Noyes Publications

INDUSTRIAL WATER TREATMENT CHEMICALS AND PROCESSES
Developments Since 1978

Edited by M.J. Collie

Chemical Technology Review No. 217
Pollution Technology Review No. 98

The more than 250 processes on which this book is based relate to various aspects of industrial water treatment. The tightening of standards in recent years for industrial effluents and, subsequently, water quality makes these processes particularly attractive.

Waters treated range from boiler water to cooling towers to process effluents and wastewaters; and the chemical agents used and methods of treatment include scale and corrosion inhibitors, flocculants, coagulants, adsorbents, biocides, flotation aids, metals removal, metals recovery, and dewatering.

The large section on wastewater treatments covers the removal of organic and inorganic substances. Many of these processes have industry-specific applications—in paper, textile, food, or chemical manufacture; others are intended for general use.

The condensed table of contents listed below gives **chapter titles and selected subtitles.** Parenthetic numbers indicate the number of processes per topic.

1. SCALE AND CORROSION INHIBITORS (59)
 Scale Inhibitors
  Amino-Phosphonic-Sulfonic Acids
  Acrylic Copolymer Compositions
  Synergistic Blend for In-Service Scale Removal in Heat Transfer Systems
 Corrosion Inhibitors
  Hydroquinone for Boiler Feed Water Systems
 Scale and Corrosion Inhibitors
  Aminoalkylenephosphonic Acid plus Carboxylic Acid and Molybdate
 Scale and Corrosion Inhibitors/ Dispersants

2. FLOCCULANTS, COAGULANTS AND ADSORBENTS (43)
 Flocculants
  Polycondensates of Epihalohydrins and Amines
  Polyquaternary Ammonium Compound
 Coagulants
  Adsorbents and Adsorbent Processes
  Aluminum Oxide Treated to Delay Crystallization

3. BIOCIDES (25)
 Disinfecting Wastewater
 Activated Oxygen Product
 Scale or Corrosion Inhibitors/Biocidals

4. METAL REMOVAL AND RECOVERY (30)
 Metal Removal
  Catalytic Reduction of Chromate
  Highly Selective Mercury Adsorbent
 Metal Recovery
  Bubble Fractionation Process for Copper Ions
 Use of Complexing and Sequestering Agents

5. DEWATERING COAL AND MINERAL SLURRIES (13)
 Dewatering Mineral Suspensions
  Hydrophobic Alcohol plus Nonionic Surfactant
 Dewatering Coal Slurries

6. FLOTATION AIDS AND PROCESSES (11)
 Flotation Processes
  Stepwise Introduction of Gas
  Zirconium Compounds as Flotation Aids

7. WASTEWATER TREATMENT (77)
 Removing Inorganic Substances
  Arsenic Removal
  Scavenging Hydrogen Sulfide
 Removing Organic Substances
  Complexing Phenolics with Acylated Polyamines
  Pressure-Hydrolytic Treatment of Effluent
  Removal of Residual Acrylonitrile Monomer
  Two-Stage Oxidation Process
 Removing Both Organic and Inorganic Substances
  Sedimentation and Thickening Zones in Same Tank
  Multilayer Filter System
  Compact Units